RETURN MIGRATION OF THE NEXT GENERATIONS

For Rebecca, Michael and Katherine,
and the 'Next Generations' – Kira, Riley and …

Return Migration of the Next Generations

21st Century Transnational Mobility

Edited by
DENNIS CONWAY
Indiana University, USA
and
ROBERT B. POTTER
University of Reading, UK

ASHGATE

Published by
Ashgate Publishing Limited
Wey Court East
Union Road
Farnham
Surrey GU9 7PT
England

Ashgate Publishing Company
Suite 420
101 Cherry Street
Burlington, VT 05401-4405
USA

www.ashgate.com

British Library Cataloguing in Publication Data
Return migration of the next generations : 21st century
transnational mobility.
1. Return migration. 2. Children of immigrants--Attitudes.
3. Alien labor--History--21st century.
I. Conway, Dennis, 1941- II. Potter, Robert B.
304.8'0905-dc22

Library of Congress Cataloging-in-Publication Data

Conway, Dennis, 1941-
Return migration of the next generations : 21st century transnational mobility / by Dennis Conway and Robert B. Potter.
p. cm.
Includes bibliographical references and index.
ISBN 978-0-7546-7373-6
1. Return migration--Case studies. 2. Transnationalism--Case studies. 3. Emigration and immigration--Case studies. I. Potter, Robert B. II. Title.

JV6101.C66 2009
304.8--dc22

2009005309

ISBN 978-0-7546-7373-6

Printed and bound in Great Britain by
TJ International Ltd, Padstow, Cornwall

Contents

List of Figures

List of Tables

Notes on Contributors

Anastasia Christou is currently Research Fellow at the Sussex Centre for Migration Research, University of Sussex, UK. She studied English Literature, Philosophy and Government (BA), International Relations and Comparative Politics (MA), obtained a Graduate Professional Certificate in International Law and Diplomacy at St John's University, New York, USA, completed a DPhil in Geography at the University of Sussex, UK and a PhD in Geography at the University of the Aegean, Greece. As a human geographer she has expertise in social and cultural geography researching within the areas of Ethnic and Migration Studies, Greek-American Studies and Modern Greek Studies, while having widely published on issues of migration and return migration; the second generation and ethnicity; space and place; transnationalism and identity; culture and memory; gender and feminism; home and belonging. Her recent book, *Narratives of Place, Culture and Identity: Second-Generation Greek-Americans Return 'Home'* (Amsterdam University Press, 2006) draws on all these issues.

John Connell was educated in Leeds and then at University College London (BA, PhD). He worked at the Institute of Development Studies, University of Sussex and then the Australian National University (Canberra). While there he spent 18 months on the island of Bougainville (Papua New Guinea) studying long-term social and economic change. He became a Lecturer at Sydney University in 1977 and eventually Professor of Geography in 2000. He has published widely on the Pacific region and on migration. The most recent of these are *Environment, Development and Change in Rural Asia-Pacific* (Routledge, 2006, edited with Eric Waddell), *Tourism at the Grassroots. Villages and Visitors in the Asia-Pacific* (Routledge, 2008, edited with Barbara Rugendyke) and *The Global Health Care Chain* (Routledge, 2009). Most of his publications have covered development issues in island states, but he has also written two other books; one on the geography of music and the other on Sydney as a global city. In between these writing projects he still manages to be a rather inadequate goalkeeper in the Eastern Suburbs Over-35s Football League.

Dennis Conway was educated at Cambridge (BA) and the University of Texas at Austin (MA, PhD). He has published widely in four related areas – migration, urbanization, tourism, development and environmental sustainability – with the Caribbean and Nepal as regional specializations. He joined Indiana University in 1976, retired in December 2005 and is now Professor *emeritus* of Geography at Indiana University, Bloomington, Indiana. He has written over

100 journal articles and book chapters, to date. His recent books include *The Experience of Return Migration: Caribbean Perspectives* (with Robert Potter and Joan Phillips 2005); and *The Contemporary Caribbean* (with Robert Potter, David Barker and Thomas Klak 2004); and *Globalization's Contradictions: Geographies of Discipline, Destruction and Transformation* (with Nik Heynen 2006). He is currently conducting research on transnational migration patterns and processes in the contemporary Caribbean (with Robert Potter [Reading] and Godfrey St Bernard [UWI-St Augustine]); specifically focusing on youthful returning professional's migrant experiences in Trinidad and Tobago.

Graeme Hugo is University Professorial Research Fellow, Professor of the Department of Geographical and Environmental Studies and Director of the National Centre for Social Applications of Geographic Information Systems at the University of Adelaide. His research interests are in population issues in Australia and South East Asia, especially migration. He is the author of over three hundred books, articles in scholarly journals and chapters in books, as well as a large number of conference papers and reports. In 2002 he secured an ARC Federation Fellowship over five years for his research project, 'The new paradigm of international migration to and from Australia: dimensions, causes and implications'.

Eunice Akemi Ishikawa is an Associate Professor of Sociology in the Department of International Studies at the Shizuoka University of Art and Culture, Hamamatsu city, Shizuoka prefecture, Japan. As of November 2008, she has taught courses such as Ethnicity, International Migration and the second and highest level of the institution's Portuguese language. In addition to her teaching duties, she is conducting research related to the return migration of Brazilians of Japanese ancestry to Japan and, in addition, is contributing to a colleague's gender studies research. A native of Londrina, in the state of Paraná, Brazil, Ms. Ishikawa earned her BA in International Relations from Tsukuba University, Japan, and MA in International Studies from Tokyo University of Foreign Studies, Japan. She is currently pursuing her Ph.D. through Ochanomizu University, Japan.

Audrey Kobayashi is a Professor and Queen's Research Chair in the department of Geography at Queen's University, Kingston, Ontario. Her PhD dissertation (1983) at the University of California Los Angeles examined the impact of emigration to Canada in a turn-of-the-century Japanese village. Her research interests have remained focused on immigration issues, incorporating questions of human rights, anti-racism, gender, employment equity and social and spatial justice, all topics on which she has published extensively. She is also involved in a number of community-based projects, working with municipalities and NGOs to create barrier-free environments for immigrants, people of colour, and persons with disabilities.

Helen Lee is a Senior Lecturer in Anthropology in the School of Social Sciences at La Trobe University, Melbourne, Australia. Since the 1980s, her research has focused on the people of Tonga, both in their home islands in the South Pacific and in the diaspora. Her doctoral research on Tongan childhood was published in 1996 as *Becoming Tongan: An Ethnography of Childhood* (Helen Morton, University of Hawai'i Press). Helen's more recent work has been with the Tongan diaspora, on which she has published numerous articles and the book *Tongans Overseas: Between Two Shores* (Helen Morton Lee, 2003, University of Hawai'i Press). Her work on the Tongan diaspora is based on both conventional ethnographic fieldwork and 'cyber-ethnography', including a study of Tongan web-based discussion forums. Her current research focuses on second-generation Tongan transnationalism.

David Ley is Canada Research Chair of Geography at the University of British Columbia in Vancouver. His research and teaching emphasize the social geography of large cities, particularly changing socio-spatial and cultural relations in inner city neighbourhoods. Earlier research analyzed gentrification in post-industrial cities; current work examines immigration and urbanization. Books include *The New Middle Class and the Re-making of the Central City, Neighbourhood Organizations and the Welfare State* (with Shlomo Hasson), *A Social Geography of the City,* and *Millionaire Migrants,* a forthcoming study of wealthy Chinese transmigrants to Canada, to be published by Blackwell. He is a Fellow of the Royal Society of Canada and of the Pierre Trudeau Foundation.

Cluny Macpherson is Professor of Sociology at Massey University's Auckland Campus, New Zealand. He has been studying Pacific migration and settlement in sending and receiving societies since 1970 and was co-author of the first comprehensive social survey of the Samoan population in New Zealand, *Emerging Pluralism: the Samoan Community in New Zealand* (Longman Paul 1973). He has been active in both family and village affairs in New Zealand and Samoa since 1970 and divides his time between Samoa and New Zealand.

La'avasa Macpherson is currently a Research Associate with Oceania Inc. in Auckland, New Zealand. She has worked on a range of Samoan research projects from the economic and social outcomes of out-migration, to Samoan ethno-medicine, and youth suicide She has been active in both family and village affairs in New Zealand and Samoa since 1960 and divides her time between Samoa and New Zealand. She is co-author, with Cluny Macpherson, of *Samoan Medical Belief and Practice* (Auckland University Press 1990) and of *The Warm Winds of Change in the Pacific* to be published by Auckland University Press in 2009.

Joan Phillips is currently a Research Fellow at the Policy Studies Institute, London. She is a Barbadian national and gained BA and MPhil degrees in Sociology at the Cave Hill Campus of the University of West Indies in Barbados, before completing her PhD at Luton University, UK. Her research interests include transnational return migration to the Caribbean, asylum seekers and refugees and sex tourism.

Robert B. Potter is currently Head of the School of Human and Environmental Sciences and Professor of Human Geography at the University of Reading, UK. He has conducted research on urbanization, planning and development in the Caribbean region since 1980 and since 1999–2000 he has been investigating second-generation transnational migration to the Caribbean, focusing on Barbados, St Lucia and Trinidad and Tobago. Several recent books, published articles and articles in press have resulted from this work, which has been funded by the British Academy, the Leverhulme Trust, National Geographic Society (with Dennis Conway and Godfrey St Bernard) and the Institute of Advanced Study, Indiana University.

Godfrey St Bernard obtained his PhD in Social Demography from the University of Western Ontario, London, Ontario, Canada in 1993 and is currently a Fellow of the Sir Arthur Lewis Institute of Social and Economic Studies (SALISES – formerly the Institute of Social and Economic Research (ISER) in the University of the West Indies, St Augustine Campus, Trinidad and Tobago. He has been the beneficiary of several prestigious fellowships and awards, including the CIDA/CELADE Fellowship in 1987 and 1989 and a Population Council Fellowship in 1991. His publications cover a range of issues including population dynamics, return migration, development studies, Caribbean family structures, adult literacy, social measurement, ethnicity and youth. He is also the co-author of a book entitled: *Behind the Bridge – Politics, Power and Patronage in Laventille, Trinidad.* In the 1980s, he worked as a professional statistician in the Government of Trinidad and Tobago and continues to be strong advocate for the development of official statistical systems. His current research interests include social measurement, population dynamics in Caribbean societies and human development challenges.

Preface

The idea for this book emanated from the success we had in publishing *The Experience of Return Migration* in 2005, along with Joan Phillips. Indeed, it was 'a first of its kind', being a collection of original works about pre-retirement return migration to the insular Caribbean. The contributions included in that regional-oriented collection focused on topics such as return visiting, circulation, temporary return sojourns, as well as the more permanent return migrations of younger pre-retirement cohorts and their adjustments and adaptation experiences to life 'back home'. These young returnees appeared to be of a different order and overall calibre when compared side-by-side with the return of the initial cohort of emigrants – the 'Windrush generation' – who had moved away from their West Indian/Caribbean homes in the aftermath of World War II and who were returning home on retirement in their late 50s, 60s and 70s. These elderly returning retirees were typecast as a cohort of 'first-generation' emigrants, and while those who eventually returned were never a majority (with more staying 'over there', for good, or for the good of their children), their contributions 'back home' were rarely considered to be influential.

In other Caribbean islands that did not have colonial ties to the UK, the initiation of moves off the island either coincided with the post-World War II exodus from the Commonwealth Caribbean, or followed it soon thereafter. For example, the emigration and circulation of Puerto Rico's 'first-generations' from the dependent territory of Puerto Rico to America began in response to *Operation Bootstrap's* considerable restructuring in the mid-1950s and continued apace through the 1970s. And, the Next Generations would follow their predecessors so that to this day, temporary circulation and moving back and forth between Puerto Rico and the US mainland has become an entrenched feature of Puerto Rican life as a 'commuter nation'.

Accordingly, the overseas-born children of the 'first-generation' Caribbean emigrants were labelled the 'second-generation', and the conventional wisdom in the metropolitan host societies was that they would assimilate more easily than their immigrant parents and gradually become accustomed to life and livelihoods in America, Canada, France, Britain, or wherever they were born and grew up. Surprisingly and counter-intuitively, over the last decade or two, some second-generations have been returning to their ancestral homelands in their mid-30s and 40s. Also among these new return flows or more temporary circulations, were the children who had been born in the Caribbean, but whom had been taken overseas by their emigrating parent(s), duly labelled the 1.5-generation. They too were expected to assimilate into their new host societies and become new immigrants.

In our enquiries, the editor-authors of this collection have sometimes found it necessary to challenge conventional wisdoms concerning processes of Caribbean migration and development that seem to have prevailed beyond their usefulness, when the editors' Caribbean field experience and information seem to point to alternative interpretations. This was very much the case with examinations of the return experiences of second-generation 'Bajan-Brits', and Trinidadian 'prolonged sojourners'. We found ourselves countering commonly held views on Caribbean return migrants as being at the end of their working lives and thus not being likely to serve as 'agents of change'. The notion that remittances from the diaspora were the latest phase of continuing dependency on external forces and effectively 'crumbs from the tables of the rich' was also challenged. Rather such influxes were re-assessed as valuable providers of social and economic capital. A further criticism and call for re-evaluation surrounded early opinions that the post-World War II emigration of West Indian/Caribbean first-generation migrants had contributed to irreplaceable and severe 'brain drains'. Though this might have been the case at the outset, the resultant diasporas, the deepening development of transnational networks and the growth of transnational circulation as an alternative strategy, have all led to changes in the system dynamics of some island societies, to the point where 'brain circulations' or the 're-draining of the brains' are now occurring.

We found we were examining new forms of migration and circulation that no longer conform to the old rules. Furthermore, and with reference to several of our Caribbean-based projects, we were dealing with modernizing peripheral societies in which overseas experience, participation in higher education abroad, and transnational networks of mobility, exchange, communication and knowledge were no longer the exception, but were becoming more and more the prevailing norms. The *next generations* were not all emigrating. Ties to their parent's homelands remained strong, dual- and multiple-identities proliferated and transnational lives and livelihoods 'across borders' were appearing to be realized as part of their privileges, rather than being beset by obstacles and disadvantages. Return migration and circulation were being considered as worthwhile mobility alternatives for some among the *next generations.* Yet, surprisingly, these new realities of 21st century migration have generally remained beneath the radar of intensive research.

Accordingly, when we look to the future, the second and 1.5 generations will be followed, logically, by the third and the 2.5 generations, with intra-generational influences on the succeeding migration and circulation options quite possible, so that expanding the optic to examine the *Return of the Next Generations* becomes an important strategy. Thus, we felt that we had a sufficiently worthwhile core of materials to warrant putting together a wider collection of essays on the experiences of the *next generations* of return migrants, in which both the contexts and the first-generations' successors – the youthful, pre-retirement *next generations* – would be our particular focus. As a result, we invited a group of active scholars to report on their research in the field, so that we could provide a more global perspective on this relatively new and under-researched aspect of 21st century migration. We hope that the results contained in this volume will prove valuable for all those working on the new realities of 21st century migration. In terms of our own research agenda, we have already begun formulating the new questions that will likely form its substantive empirical basis.

As editors, we should like to acknowledge and thank sincerely those bodies that have funded our research on these topics over the past few years. These include the Leverhulme Trust, the National Geographic Society Committee for Research and Exploration (NGS-CRE) and Indiana University's Institute for Advanced Study. It is their funding that has made this volume a realizable goal.

Dennis Conway and Robert B. Potter
December 2008

Chapter 1

Return of the Next Generations: Transnational Migration and Development in the 21st Century

Dennis Conway and Robert B. Potter

In the global and transnational world of the first decade of the 21st century, people, capital, information, and cultural traits are circulating globally at increasing rates. International circuits of migration are no longer bi-polar and more of them are developing as multi-local transnational networks of movement. The resultant diasporas are far flung and multi-nodal and the cross-border social spaces that are being formed are serving to define, more and more, the social worlds of those involved. Such networks, once established, are themselves serving as facilitators of new waves of migration, of return, moving on and re-return. Indeed, the growing complexity of contemporary international migration pathways, nodes, moorings and way-stations, as well as their wider global reach, is without precedence.

To this point, extant scholarship on international return migration has commonly focused on elderly, first-generation retirees who have consummated their long-held desires and intentions to return to the global South homelands from which they emigrated in their youth. During the post-World War II era, there were appreciable outflows and emigrations of colonial and post-colonial 'hopefuls' who sought better lives abroad, better educational opportunities for themselves and their children, or the better opportunities that they were initially denied in their global South source societies. They left behind impoverishment or inequality of opportunity, impenetrable glass ceilings because of gender, race or educational shortcomings, and a dearth of opportunities in their peripheral homelands. In going abroad, they contributed to the rebuilding and reconstruction of war-torn economies, joining the staff of rapidly enlarging public service sectors in Europe and North America and working as 'replacement' labour in the modernizing metropolitan economies of their respective colonial 'mother countries' or neocolonial partners. They set off to seek their fortunes, conducting 'rights of passage' as inquisitive and adventurous youth, though future plans differed in respect of whether they left with an intention to return, to perhaps return when they had become successful, to return on retirement, or indeed, never to return.

These first generations did not always emigrate in waves, nor did they all go abroad at the same time, because differing immigration legislation and policies in the host destinations influenced entry timing. Equally, different conditions in sending societies across the global South – peripheral nations, islands, dependent colonies, and neocolonial or subordinate, occupied or administered territories – obviously served as instigating environments. Recruitment efforts in specific countries, also played their part, so that the first-generation's experiences would always be varied, in relation to the contexts of the sending and receiving societies – the 'welcome' or 'unwelcome' messages host societies portrayed and the workplace and societal niches such new immigrants found themselves in, climbed out of, or in which they remained. We can generalize, however, that regardless of the exact timing of the initial sortie, the resultant emigration of the first generations began the international, and later, transnational migration 'cycles and circuits' of the families and generations that followed. The resultant transnational diasporas would remain connected as networks consolidated, as intra-generational mobility became more possible, and as circuits of people, knowledge and resources became commonly-practiced exchanges between multi-local nodes in ever-widening, spatially-diversified social fields 'across borders' (Conway 2007). At the same time, the life courses, further migrations and circulations of the first generations would play out, while remaining an influential part of the *next generations*' migration calculations.

On retirement, in preparation for retirement, or on the death of their partner, many (though far from the majority) among the first-generations left the metropolitan global North destinations to which they had emigrated and returned to their ancestral homelands to live out their days. Others might have intended to return on retirement, but their children's assimilation into their adopted new home, and unwillingness to accompany them, changed their minds. Some others will have just procrastinated until it was too late, so that they were too firmly settled in their new home to view return on retirement as a viable option. Ageing and infirmity might also intervene, with health service provision becoming extremely influential in the decision to stay in the metropolitan, host destination. So, many of these first generation 'new immigrants' stayed away 'for good', their temporary stay or sojourn became a permanent relocation, their children and grandchildren transferred national allegiances, and became assimilated, while others – usually a comparative minority – returned on retirement. Shorter-term stays abroad and pre-retirement returns were often characterized as 'failed-migrations', or impromptu returns because of adjustment difficulties – employment disappointments, failure to reach target earnings, dropping out of school, and often scarcely-hidden racist and anti-immigrant opposition and hostility in the 'host societies' they had dared to enter with the cherished hope to succeed that characterizes such new immigrants.

The first-generations' children and grandchildren are the *next generations* who until recently have remained *relatively invisible* as a return migration cohort. They are the second generation (born overseas to an emigrant parent or parents), one-and-a-half (1.5) generation (those born at home, but who emigrated with their parent(s) at an early age), or other relatively young or youthful pre-retirement 'returning nationals', who followed the paths of the earliest wave(s) of emigrants to metropolitan North America, Europe and beyond (Asia, Oceania global cities, for example) from their global South countries of origin, but who then returned after lengthy sojourns overseas. In a few cases, where the first-generation's emigration or temporary labour recruitment occurred considerably earlier in the nineteenth century, the returning second generation will be parents who do or do not return with their 'third generation' children, so second- and third- generations' lives and livelihoods factor strongly in the return and re-return calculations. As Chapter 4 demonstrates, the Japanese-Brazilians, who were encouraged to return to Japan from Brazil in the 1980s, are an example of this situation. In other cases, second-generation parents often have their third-generation young children's welfare and education very much in mind, but most of the latter dependents will be returning as 'involuntary' movers, where their intentions and wishes concerning such moves are not likely to be as influential as their parents' motives and decisions on their behalf. Lee's examination of Tongan youths' return assessments in Australia reflects this, and other examinations of family return among Trinidadian and Australian returnees in Chapters 9 and 10 respectively, also allude to the immediate family's mobility decisions, which concern the future of the third generation. Though they are not a particular cohort of interest in this collection, what we can assert, however, is that as time passes, the third generation's return will soon factor into the return, re-return or stay conundrum as a new, and significant 'next generation'.

Much more on the relatively recent 'discovery' and examination of young pre-retirement return migrants throughout the Caribbean region is provided in this collection's concluding Chapter 11. This summary of the collection's work refers back to the authors' predecessor of this collection, *The Experience of Return Migration: Caribbean Perspectives*, which was co-edited with Joan Phillips and published in 2005, and the account starts with a synopsis of initial findings as a point of departure for this more global treatment. Chapter 11 then proposes several general themes that emerge from comparisons of the transnational and return migration experiences and practices detailed in this collection, namely, that geographical contexts, age and life-course stages, family demographics and experiences, and transnational mobility are influential 'determining' characteristics of the new circuits of return migration in the 21st century.

An exception to the general observations offered above, might be the special case of young and youthful deportees, who have been forcibly returned, often against their will and in ever increasing numbers, by US Homeland Security immigration officials and judges. Many deportees had been arrested and charged with criminal offences, served their jail sentences and have then been promptly sent back. Other deportees had been arrested and when their immigration status or work status was found to be illegal or undocumented, they too were deported promptly without much recourse to further legal procedures or hearings. Deportations in Europe of failed asylum-seekers, who then stayed on and worked in the burgeoning informal sectors as irregular migrants constitute a similar exception, though their numbers are comparatively small by comparison to the US deportation surge. Both of these forced repatriation flows have warranted attention and critical commentary in global and local media, over the internet and in policy circles; both in the deporting countries – the US, Australia and European Union members – and in the returnees' countries of origin. Except for one case – a small sample of young Tongan deportees in Chapter 3 – the special experiences of this particular group of return migrants are not featured in this collection. In large part this is due to the impromptu and coercive nature of their return migration experience, and also because their return was institutionally managed and executed. It was, therefore, rarely their own decision, not voluntary planned nor accomplished with forethought and with sufficient resources to assure its success, socially and economically.

In many cases, the invisibility of the Next Generations who voluntary return is due to the relatively small numbers involved in comparison to the larger and more familiar emigration outflows that emanate from countries of the global South and which flow into the metropolitan countries of the global North. On the other hand, one country Australia appears to stand out as an exception in terms of its relatively large number of young and youthful pre-retirement returnees. Australia also stands apart because of its government's exceptional performance in collecting comprehensive data on the country's emigration and return migration patterns. As is shown in Chapter 10, the coverage of Australia's diaspora, emigration and return migration sub-populations is without parallel, making both the country's transnational migration patterns and the data collection protocols that have been instigated an exemplary case to be imitated and replicated by other countries seeking to better manage (and measure) their international migration systems.

Some among the Next Generations are Returning

The *next generations* of young and youthful return migrants who are in our spotlight are returning in the early- to mid-career phases of their life-paths. Many are in the early family-formation stages, and their active participation in the social, cultural, political and economic 'spaces' of their ancestral homelands is the rule not the exception. Many have 'first-generation' transnational migrant parents, or a parent, who to varying degrees have initiated them to transnational practices and experiences. Many have transnational experience and worldly backgrounds, with or without direct parental guidance, help and encouragement, so that their return is part of their strategically flexible decision-making which is concerned with operationalizing their social, economic and mobility options to their own advantage, or that of their dependent children – the third generation. Many are also 'agents of change' by intention and persuasion, by example and exploits, and by *praxis*.

Looking to the future, as first-generation migrant cohorts reach retirement age and beyond, the second (and possibly third) generations are poised to become an ever more important group, likely serving as transnational cohorts, who will forge further links between the peripheral homelands in the global South and the metropolitan global North (and emerging Southeast Asia, for example). In addition, there are a growing number of young and youthful transnational migrants returning 'home' in their 30s and 40s, who left the global South in their late teens and 20s to acquire higher education, added skills and professional experience abroad. These 'prolonged sojourners' (Chapter 9) or 'returning youthful nationals' (Chapter 10) grew up and acquired secondary education in their homelands, then sought advancement and better opportunities via international circulation, but are now, like the 'one-and-a-half (1.5)' and second generations, returning while still in the prime of their working life – in mid-career and mid-family formation stages. A fourth cohort of youthful returnees originates from the global recruitment of relatively young and youthful circulating professionals and skilled workers, who undertake limited-term, contract work in the global North and global South, or other transnational fields of economic opportunity – the skilled employment sectors of health services, resource exploration, extraction and delivery, corporate commerce, trade and international communications. Again, the return flows of these global skilled and professional workers, albeit selective and numerically small in most cases, deserves our attention for their potential to serve as 'brain gains', or 'brain circulations', which go some way to offset the 'brain drain', of which they were part when they were recruited, or enticed, away.

Notably, all of these cohorts of young returnees, many or most of whom have acquired higher education, professional skills, business acumen, IT and computing experiences while abroad, have more information at their disposal via the internet, telephone and other global networked systems. And, as a direct consequence, they are more directly aware of the opportunities that are open to them, and also perhaps to the adjustments they may have to make in their new environments. They are likely to be more skilled and better endowed with stocks of social and cultural capital than their more elderly returnee counterparts were in the past. Some might very well be moving back and forth, trying out strategies to see if they can enhance their standards of living and the quality of their lives. Some may be content enough, or adventurous enough to experience a new, global life and livelihood 'between two or more worlds'. Some may re-return, after adjustment problems overwhelm their return intentions.

By means of comparative studies set in the context of new global migration experiences, the chapters of this volume ponder vital issues such as the socio-cultural adjustment faced by these transnational return migrants and the degree to which they are able to settle in, establish a satisfactory life for themselves, their partners and dependent children both in work and society, renew ties to family and contribute to local and/or regional development. How are the transnational networks of these returnees contributing to their adjustments; are they changing their situations, making them more mobile or more settled in their new homes? Are they living multi-local transnational lives to the extent that they retain multiple identities and view their global life-world as part and parcel of their new status as strategically flexible, global citizens? Are they the 'agents of change' that their retiring parents were unable to be? Are they contributing to societal development or helping mobilize changes in the workplace? How permanent are their return intentions? If they are truly transnational, and living between two or more worlds, what might be their influences and experiences 'back home'? Have some, many, or only a few, re-returned disappointed and disenchanted? How difficult are their social and economic adjustments? How influential can such small cohorts be in terms of guiding societal development, or contributing to development? Are the health, modernization, and expanding diversity of local economies and availability of wider sets of economic opportunities for such skilled professionals, important contextual factors in the effectiveness of these returnees' contributions? Is it the collective power of return migrants' agency or institutional, structural factors in their home societies that determine outcomes in the challenges return migrants take on to contribute positively and achieve their goals? These are some of the unknowns, representing questions that need to be answered.

Global Perspectives: Transnationalism and Return

The chapters that follow convey research by leading authors in the field from the UK, USA, Australia, Canada, the Caribbean and New Zealand. Many of the contributing authors are well-known 'first-generation' migration scholars, while the others are 'up-and-coming scholars', who rely on fieldwork and micro-level investigations to inform their theoretical conclusions. All the authors are 'internationalists' in their scholarly perspectives and academic agendas. The original chapters[1] commissioned for this collection are all built upon the contributors' most recent field investigations and inquiries,

1 Chapter 7, by David Ley and Audrey Kobayashi, is an exception to this claim. It was first published as a journal article in *Global Networks* (2005), but its theme and empirical evidence of transnational sojourning among Canadian-Hong Kong professionals is so well-integrated with that of other chapters that its contribution to the collection is valuable in every respect.

yet over half of these contributions are informed by decades of internal and international migration research. The authors talk 'across disciplines' – predominantly between geography, anthropology and sociology, – so, the topic is interdisciplinary, the empirical methodologies utilized are a mix of quantitative and qualitative analyses, and generalizations of relevance to social science are an attainable goal.

Return migration has recently been 're-discovered' as a significant emerging dimension of today's global labour patterns (King 2000). It is no longer the forgotten or overlooked dimension in international circuits of emigration and immigration. It is no longer disparaged as the mobility response of the unsuccessful, the failed immigrant, or the retired. Going beyond examinations of the return of first-generation, elderly, international migrants whose return to their source communities on retirement has dominated the literature to date, the present collection of essays expands the research frontier into the realms of the one-and-a half and second generations and accompanying new cohorts of youthful global contract workers and 'prolonged sojourners', and examines *their return and its consequences*.

Transnationalism, however, builds global networks in which return, circulation and other temporary or more permanent strategies of migrants and their family members, including the next generations, are highly salient components. Choosing to return to the home society of their parents then becomes an option, an opportunity, or an experiment, an initial 'testing of the waters', or a 'try-out'. The return of youthful migrants better endowed with stocks of human social and cultural capital than previous cohorts, and the flows of remittances that also form an essential part of the circuits of transnational exchanges, together constitute a global re-alignment of human and capital resources that promises much in terms of the global South's 'development future'. *Return Migration of the Next Generations*, therefore, deserves our attention, as much as the next generations' adaptation and assimilation experiences in the overseas, metropolitan destinations in which they were born, or in which they grew up, were educated and acculturated.

Given the newness of the phenomenon under examination, its emergence as a significant interdisciplinary topic in international studies, global migration and development studies and the uniqueness of this volume's focus on the *return migration of the next generations*, this collection is sure to act as a comparative template for further research in wider global contexts. Our book is about transnational linkages that foster the return of young and youthful cohorts, and the UK, USA, Japan, Australia, southern Europe feature prominently, along with Caribbean and Pacific island societies. Global interconnectedness is going to continue to deepen, and transnational circuits and networks are not only going to strengthen and deepen, but the multi-local character of these global webs is going to metamorphose into new, non-traditional forms. Return and temporary circulation, global skilled-labour recruitment, short-term contract work for foreign professionals, global circulations of students, young professionals, artisans, athletes and entertainers are going to be permanent features of all future global, transnational networks and diasporas. 'Brain circulations' might be a collective term, which re-situates the return circuit as more than failed migration, or a loss of migrant human capital. Our collection starts the proverbial ball rolling in this regard.

The Experiences of Second-Generation Return Migrants

This first group of chapters in this volume explicitly targets the return experiences of relatively youthful second-generation migrants, who have returned to the ancestral homes of their forebears. With no order of importance implied, Samoa, Tonga, Japan, Barbados and Greece are the return destinations, while the metropolitan host countries from which they have predominantly returned are New Zealand, Australia, Brazil, the UK and the USA, respectively.

In Chapter 2, Cluny MacPherson and La'avasa Macpherson deal with New Zealand-born Samoans who have returned to their parents' homeland for a variety of reasons, and these diverse causes serve very much to shape their expectations of, and responses to, Samoa. At least six groups of returnees are identified: family care and service providers, cultural heritage seekers, social idealists, professionals and entrepreneurs, and explorers. 'Family care and service providers' fulfil family obligations to parents and grandparents, and return to Samoa to support or care for relatives who have retired to Samoa, or whom have stayed behind or never left. A strong cultural tradition in Samoan society that emphasizes the duty of children to those 'who raised them', including parents and grandparents, is behind this return motivation. 'Cultural heritage seekers' on the other hand, were motivated to return to improve their language fluency and to get to know the homeland better, so that they would be more fully informed about their 'authentic' Samoan identity. 'Social idealists' are returning to help in the transformation of various communal and societal aspects of the parents' homeland and 'to improve things'. 'Professionals and entrepreneurs' return to their ancestral homeland in the hope of taking advantage of opportunities which they believe are available in Samoa's small, yet modernizing economy. Lastly, 'explorers' visit the homeland of their parent's stories in search of the people, places and experiences at the heart of the tales of the homeland about which they heard as children. They too, like the cultural heritage seekers' are concerned with strengthening their Samoan identity. As the MacPhersons' demonstrate from their field work among returning Samoans, these six different groups of youthful returnees have contrasting experiences of, and responses to, Samoa. Within each group there is a range of experiences that are shaped by their understanding of the culture and language of the country and by how realistic their expectations of Samoa were before they left New Zealand. Furthermore, often the lack of cultural awareness and sensitivity among these returning youth strongly influences their ability to cope with challenges which the return move – whether temporary or relatively permanent – poses to their personal identity and their social adjustment.

Helen Lee's Chapter 3 examines the considerable ambivalence of return among second-generation Tongan-Australian youth who return, or are temporarily returned by their parents, to their parents' ancestral island home from Australia. For many of these children of Tongan migrants, travel to their parents' homeland whether for short or long stays is often a disconcerting and problematic, experience. While a few return with entrepreneurial ambitions and a desire to reap financial rewards, most are motivated more by a desire to discover their heritage and cultural identity, and an altruistic wish to 'help' what they perceive as a poor, disadvantaged country. Still others return simply because they are forced to: they are the *tipoti*, or deportees, and growing numbers of these mostly young male returnees are arriving in Tonga and attempting to find ways to cope with their new circumstances. Tongans who have remained in the islands often assume the returnees have wealth and may use their kin connections to seek 'loans' and favours. Yet the 'stay-at-homes' also often regard these young returnees with considerable resentment and even hostility, refusing to acknowledge them as 'real' Tongans and treating them, at least at first, as outsiders. In addition, returnees have to deal with the reverse, culture-shock of living in Tonga, where many of the services and facilities they took for granted in Australia are either non-existent or of comparatively poor standard.

In Chapter 4, Eunice Akemi Ishikawa examines the post-1980s return of relatively young Japanese-Brazilian families, the second and third generations, as temporary contract labourers along with third and fourth generations, their children. She specifically focuses upon the workplace experiences of these Next Generations of parents, and their children's schooling experiences as they struggle with the temporary status that characterizes them as 'outsiders' or foreigners in Japanese society, despite their Japanese ancestry which allowed them to enter. This latest influx of Japanese-Brazilians responded to recruitment at the end of the 1980s and was comprised of contractual laborers (and their families) who moved to Japan for low-skilled manufacturing jobs and the higher wages they would garner. The problems this group of returnees faced and continues to face, however, is that they are on temporary contracts and many (or most) intend to re-return to Brazil. Their children (the third, 2.5, fourth and 3.5 generations) find themselves caught between two cultures, that of their Brazilian parents and the host culture of Japan. The Japanese-Brazilian children raised in Japan speak primarily Japanese and are more familiar with the Japanese culture than their parents who, in Japan, try to retain their Brazilian cultural ethnic identity and traditions in order to lead their daily lives. Compared to their Japanese-Brazilian parents and other adults living and working in Japan on temporary contracts, the children learn the Japanese language much more readily and quickly. Thus, Chapter 4 focuses on the experiences of second- and third- generation Japanese-Brazilians who went to Japan as adults in the post-1980s period, as well as their third and fourth generation children born or raised in Japan – the latter being 2.5 and 3.5 generation children born in Brazil who accompanied their parents while babies or young children. The author's qualitative method of research for this chapter is based on extensive oral interviews with Japanese-Brazilian families, both in Japan and Brazil.

Rob Potter and Joan Phillips' Chapter 5 deals with second-generation Barbadians, or 'Bajan-Brits', who have decided to return to the birthplace of their parents, focusing in particular on patterns of socio-cultural adjustment on their return as relatively youthful 'citizens by descent'. The cumulative stages of this research project are also introduced, to demonstrate how the initial operational problems were overcome when faced with identifying such an invisible cohort of returning nationals. The remaining empirical section draws upon fifty-two in-depth qualitative interviews that were conducted by the authors with second-generation return migrants to Barbados. As a group, these youthful returnees appear to be advantaged in the employment market, being perceived as professionally adept as well as being characterized by an 'Anglo-Saxon work ethic'. In addition, these 'returning nationals' experience pronounced racial affirmation on return to Barbados, their ancestral homeland – in short, they benefit from a series of positive effects which are attendant on living in a predominantly black society following their move to the Caribbean. The research also speaks of the liminal, hybrid and 'betwixt-and-between' racialized identities of the second-generation migrants. For example, the authors found that many of the informants encountered extreme problems with respect to the formation of new friendship patterns 'back home'. Friendship patterns appear to be particularly difficult for female migrants who reported serious issues relating to both workplace and sexual competition. Thirdly, in what appeared to be very much grounded within the local Barbadian socio-cultural milieux, many of the migrants reported that they were frequently referred to as being 'mad'. This 'madness trope' can be interpreted as a means of 'othering' and 'fixing' the returnees in a manner that means that they do not have to be listened to, or their advice heeded. 'In-betweenness' for some of these second-generation Bajan-Brit returnees was problematically coined as 'being in mid-air'.

Turning to Chapter 6, Anastasia Christou looks at the contemporary life histories and return experiences of two second-generation Greek-American return migrants as expressed orally and in detailed, self-reflective written journals. The two informants, who are young, middle-class, highly skilled and highly educated professionals, imaginatively construct and practically implement their return as a project of self-discovery, of identification and belonging as well as a contribution to their ancestral homeland. In response, however, they are met with antagonism and exclusion by the homeland's unwelcoming society. Such encounters become spaces of transformation in their changing (ethnic and cultural) identification and sense of belonging. The migrants' life stories become the narratives of their transnationalism, wherein questions of identity and home are renegotiated along the lines of gender, ethnicity, class and generation, although perhaps not to ever be resolved completely. Furthermore, their state of migration and displacement becomes exacerbated throughout their adjustment processes 'back home' as they confront new and 'foreign' circumstances in a modern Greece (and modernizing Athens) that are quite different from those they imagined, or had previously experienced in the distant past.

Young and Youthful Return Migrant Experiences

This second group of four chapters turns attention to the transnational practices, adaptation experiences and migration life histories of other prevailing types of relatively youthful return migrants – temporary circulating migrants, prolonged sojourners, skilled workers in the global health industry and young professionals. The chapters indicate that while some prosper, adapt and adjust and evaluate their return in positive terms, others find ambivalence, obstacles and disappointment, so the general finding concerning these *next generations*' return experiences is again varied and mixed. And the background contexts to their moves are also varied and differentially influential. Once more with no order of importance implied, Hong Kong, Niue and the Cook Islands in the South Pacific, Trinidad and Tobago and Australia are the return destinations in question. The metropolitan host countries from which these country's overseas expatriates and nationals have returned are sometimes more than one singular host, so that they are respectively, Canada (Vancouver), New Zealand, the UK and North America (Canada and the USA), Europe, the UK and East Asia.

David Ley and Audrey Kobayashi reconsider the meaning of return migration for Hong Kong-Canadian middle-class returnees from Canada in Chapter 7. Drawing upon these returnees' own stories of their transnational experiences and practices offered in focus groups in Hong Kong, they find that involvement in 'cross the Pacific' movement between Hong Kong and Vancouver, Canada is undertaken strategically at different stages of the family life course. For such professionals, return trips to Hong Kong typically occur for economic reasons at the stage of early or mid-career. A second move to Canada may occur later with teenage children for education purposes, and even more likely is migration at retirement when the quality of life in Canada becomes a renewed priority. Strategic switching between an economic pole in Hong Kong and a quality of life pole in Canada identifies each node or 'mooring' as a separate 'space' within an extended but unified transnational social field. Return as 'transnational sojourning' is a temporary move back to Hong Kong for these mobile transnationals and a re-return move back to Canada completes the circulation.

John Connell's Chapter 8 traces the return migration of skilled health workers, mainly nurses, in two small Pacific island states, Niue and the Cook Islands that are dependent territories with strong ties to New Zealand. This case study is set within the wider regional context, in which the emigration of health workers appears to be a dramatic and significant example of 'brain drain' from many of these South Pacific islands. In Niue and the Cook Islands, on the other hand, some do return, though not necessarily permanently. Such returns are potentially a 'brain gain' to partially offset losses in this crucial service sector. Their impact could be significant for both social and economic development, in terms of gains to health services (through new skills and wider experience) and to the economy for their (partial) investment of overseas-generated incomes. Yet many are dissatisfied (for example, with the pace of life, hierarchical structures and their own inability to implement change). For these disappointed returnees, such moves 'back home' may be temporary rather than permanent and this appears to be the case in Niue, where opportunities are relatively few. Return migration, particularly of such skilled health workers may thus be more problematic than for South Pacific island returnees with other professional backgrounds and entrepreneurial skills.

In Chapter 9, Dennis Conway, Rob Potter and Godfrey St Bernard examine the transnational experiences and practices of a self-selective cohort of young and youthful Trinidadian transnational migrants of working age, who have decided to give it a try 'back home'. These are a group of returnees who had left Trinidad as school-leavers for further education, skill acquisition and overseas experience, but who stayed after their studies and had thus lived overseas for more than twelve years before eventually returning. In the account, the focus is on the experiences of 25 prolonged sojourners, this latter group being long-term circulators, though still young and in pre-retirement careers and mid-family formation ages, in their 30s and 40s. Analysis of the informants' own life-stories or 'narratives' focuses on how their family and community relationships and individual and family migration decision-making influenced their emigration and return. A further focus is on how their transnational experiences as 'prolonged sojourners' prepared them for return, facilitated their adaptation and smoothed their transition 'back home'. This cohort of young returning migrants is characterized by extended information available at their fingertips via the internet, telephone and other global networked systems,. Many have made frequent return visits during their extended stays abroad and as a consequence, are more directly aware of the opportunities that are open to them, and the adjustments they may have to make.

In the Trinidad context, the following questions are asked and answered: what motivated these young and youthful prolonged sojourners to emigrate in the first place and what then prompted their return at mid-career and mid-life course ages – the 30s and 40s? Do they experience different processes of adjustment in the workplace and social spheres? Do they maintain their transnational identities – duel-citizenship, property and business ties – and continue their transnational practices – such as repetitive visits and circulations – which keep their future migration options open and flexible? Do some of them find social adaptation too difficult and plan to re-migrate to the metropolitan countries they returned from? The informants' life stories serve as the empirical substantiations or qualifications surrounding these questions.

In Chapter 10, Graeme Hugo deals with the exceptional case of Australia and he addresses the important 'development question' that revolves around the human capital and demographics balance sheet that pits emigration and 'brain-drain' against return 'brain gains' and/or 'brain circulations'. Australia's comprehensive database on emigrant and return sub-populations is one distinction. The other is the size of the country's expatriate sub-population; there is a diaspora of around one million people living in foreign countries, many of whom are young and youthful, highly-skilled expatriates who retain ties to home, and have every intention of returning. Chapter 10 begins with an analysis of a unique data set maintained by the Australian Department of Immigration and Multicultural Affairs (DIMIA), which allows Graeme Hugo to trace the extent to which young Australians who move to foreign countries on a more or less permanent basis, then actually, or eventually, return. This allows him to conduct a comprehensive analysis of the scale of movement and return among young Australians and of their economic characteristics, and to provide readers with the substantiating tabular details of the country's internationally mobile population, the like of which is unavailable in any other country. Notably, the propensity to return among the expatriate 'one million more' is confirmed without question. One feature is a strong predilection for return migration to be common among the 30s and early 40s age groups of expatriates. Rather than the expected pattern that returning retirees predominate, quite a different life-course cohort appears to be involved in Australia. Overseas Australians who are at the early or middle stages of family formation and at, or near, the peak of their career development appear to be the ones most inclined to return. Their return is motivated not so much by economic and employment factors but more by lifestyle and family considerations.

In the second part of the chapter Graeme Hugo bases his analysis on a survey of returned Australians and an in-depth study of a small part of that group, which looks into the characteristics, attitudes and motivations behind return. In particular, the reasons for return are probed as well as issues of identity, attachment to Australia as well as linkages with, and commitment to, other countries. The impacts of returnees on the Australian economy and society are investigated, in particular, the extent to which they have made use of their overseas experience and linkages. In addition, the difficulties they experience in adjusting back to life in Australia are assessed.

In the concluding Chapter 11, as editors we summarize the contributions made in this collection. The account considers the return experiences of young and youthful, 'second' and 'one-and-a-half' (1.5) generations, circulating global skilled workers and professionals and new cohorts of 'transnational sojourners' and 'prolonged sojourners'. Two contributions stand somewhat apart from this general list of the first-generation's *next generations*, in large part because of their contextual *exceptionalism*. One is the chapter by Eunice Ishikawa, where second- and third-generation Japanese-Brazilian parents and their children's adaptation problems on their recruitment and return to Japan their ancestral homeland are examined (Chapter 4). The other is the chapter by Graeme Hugo, wherein he analyses Australia's comprehensive tabular data, and national survey data on that country's comparatively youthful emigrant sob-population; many of whom are returning or intend to return, both as first- or second- generations (Chapter 10). They both involve the return of the *next generations*, although broader (less-specific) characterizations are being utilized in these two derivative cases.

This caveat aside, all the contributing case studies are examined in a variety of societal contexts; many in the home societies of their first-generation parents, some in metropolitan destinations, but all in 'transnational spaces', in terms of their life-worlds. Many returnees are globally 'in-between', as their multiple identities and transnational circuits of mobility both make them flexible as well as bi- or multi-national in affiliation. While some are able to contribute to societal advances in their parents' home territories as 'agents of change', for others their experiences of adjustment are difficult and their high expectations are dashed because of gender, class and racial/ethnic issues and conflicts.

Initial theoretical generalizations can be offered based upon the comparative analyses afforded by the wider global reach of the collection's insular societies and national cases. It is anticipated that some analyses will find young returnees acting as agents of change, while others will find that disappointment is more common than success. Some returnees might be too early in their professional careers to make a real mark, others might find the conditions too difficult and insecure. Some might despair that they will never ever make a difference, and others are ready to quit and re-migrate. It is to be expected, therefore, that the full range of adjustment problems and social and cultural barriers will emerge from the contributions in our comparative collection. Such a wide range of experiences is to be expected because of the decidedly different geographical contexts of our contributors' case studies and the hemispheric differences in the transnational worlds and 'spaces' within which these young and youthful return migrants find themselves living, in an essentially 'betwixt and between' manner.

However, the Caribbean and South Pacific comparisons, when added to the other global cases – dealing with Greece, Japan, Hong Kong, Australia – do provide a useful 'starter set' of empirical research on the adaptation experiences, successes and disappointments of today's cohorts of young and youthful returnees to global South homelands that should serve as a benchmark for others to draw upon. If other countries follow the lead of Australia in terms of migration data collection and migration systems analysis, then it is likely that this relatively invisible and under-reported aspect of contemporary global migration – namely, the young returning *next generations* – will be found to be more prevalent than anticipated or expected. Returnee's stocks of human and social capital, when added to remittances and the transnational transfers of knowledge, information, capital and goods in kind and services, are seriously undervalued, in part because of a general dearth of international migration information that is specific enough to provide valid estimates and in part because simple numerical assessments (undercounts) minimize (and often trivialize) temporary and short-term circular flows, return visits, transnational sojourns, and flexible mobility.

When such national and international monitoring is in place, and better estimates of the whole range of global migratory flows are forthcoming, then it is more than likely that the initial findings of this collection will be upheld, or modified, but less likely that they will be countered. We fully anticipate validation of our general conclusion that the 'transnational presence' back home of professional returnees will be found to be more meaningful, more effective and more influential, both for the migrants (parents and children and family dependents) and the homeland societies to which they have returned in order 'to make a difference', to 'give something back', to 'make a new life' and to 'provide better livelihoods and chances for their loved ones' (Chamberlain 2006; Conway and Potter 2007; Potter and Conway 2008). In partnership with the author's earlier 2005 Caribbean collection on *The Experience of Return Migration* (Potter, Conway and Phillips 2005), this current collection on the *Return Migration of the Next Generations* not only broadens the scope and deepens our understanding of global migration's dynamism and its ever-changing patterns and processes (Conway 2006; Castles and Miller 2003), but it identifies several ways forward for further research and enquiry that should prove both interesting and fruitful to pursue.

References

Castles, S. and Miller, M.J. (2003), *The Age of Migration: International Population Movements in the Modern World*, 3rd edn, New York: Guilford.

Chamberlain, M. (2006), *Family Love in the Diaspora: Migration and the Anglo-Caribbean Experience*, New Brunswick, NJ: Transaction Publishers.

Conway, D. (2006), 'Globalization of labor: increasing complexity, more unruly', in Conway, D. and Heynen, N. (eds), *Globalization's Contradictions:*

Geographies of Discipline, Destruction and Transformation, London: Routledge, 79–94.

Conway, D. (2007), 'Caribbean transnational migration behaviour: reconceptualizing its "strategic flexibility"', *Population, Space, Place*, 13: 415–31.

Conway, D. and Potter, R.B. (2007), 'Caribbean transnational return migrants as agents of change', *Blackwell Geography Compass* [online early October 2006], 1(1): 25–45.

King, R. (2000), 'Generalizations from the history of return migration', in Ghosh, B. (ed.), *Return Migration: Journey of Hope or Despair?*, Geneva: International Organization for Migration, United Nations, 7–55.

Ley, D. and Kobayashi, A. (2005), 'Back to Hong Kong: return migration or transnational sojourn?', *Global Networks*, 5(2): 111–27.

Plaza, D.E. and Henry, F. (2006), *Returning to the Source: The Final Stage of the Caribbean Migration Circuit*, Jamaica, Barbados and Trinidad and Tobago: University of the West Indies Press.

Potter, R.B. and Conway, D. (2008), 'The development potential of Caribbean young return migrants: "making a difference back home ..."', in van Naerssen, T., Spaan, E. and Zoomers, A. (eds), *Global Migration and Development*, Abingdon, Oxford: Routledge, 213–30.

Potter, R.B., Conway, D. and Phillips, J. (2005), *The Experience of Return Migration: Caribbean Perspectives*, Aldershot and Burlington, VT: Ashgate.

PART 1
Second-Generation Return Migrant Experiences

Chapter 2
'It Was Not Quite What I Had Expected': Some Samoan Returnees' Experiences of Samoa

Cluny Macpherson and La'avasa Macpherson

Samoa is a small, independent south Pacific state.[1] Its population is around 190,000 and, as a consequence of post-World War II migration, another 200,000 people of Samoan descent live in expatriate settlements around the Pacific Rim.[2] The formation of expatriate, diasporic communities, their discrete histories and different organizational structures (Fitzgerald and Howard 1990), has been well-documented over the last 40 years. They have been profiled in New Zealand (Pitt and Macpherson 1974; Macpherson 1978; Ministry Of Pacific Island Affairs 1999); California (Shu and Satele 1977; Rolff 1978); Seattle (Kotchek 1982); Hawaii (Forster 1954; Blumbaum 1973; Franco 1987; Franco 1990); Australia (Va'a 2001); Fiji (Tuimaleai'ifano 1990) and Papua New Guinea (Sinclair 1982).

Many Samoans who left after World War ll, for social, educational and material opportunities in nations around the Pacific rim, maintained social and economic links to their families and villages in Samoa. Remittances from migrants have at times accounted for as much as 33 per cent of Samoa's GNP (Ahlburg 1991, 1995). While some early migrants deliberately severed ties with Samoa (Macpherson 1984, 1991), many visited regularly to commemorate rites of passage, to attend to family land and chiefly title matters, and for holidays with their overseas-born, second-generation children. After retirement some returned to Samoa to settle.[3] This international circulation of both first- and second-generation migrants between 'home' and 'expatriate' communities has contributed to a continuing and relatively high level of awareness of, and engagement with, developments in Samoa among expatriate Samoans.

Since 1990, Samoa has achieved some of the highest rates of social and economic development in the Pacific region. Many gains have come from

1 It is located at 13° south latitude and 174° west longitude.

2 The largest numbers live in New Zealand, but there are also significant numbers in Australia, USA, and Fiji.

3 This became more attractive when New Zealand made its old age pensions portable to make it possible for Pacific retirees to return to their homes.

macro-economic and governance reforms promoted by international financial institutions (World Bank 1993) and embraced by Samoan governments since the early 1990s (Government of Western Samoa 1996; Government of Samoa 1998; Government of Samoa 2000). Macro-economic reforms have produced increases in efficiency and productivity. Governance reforms have resulted in more transparency and accountability in the public sector. While further gains may be extracted from ongoing reform of the macro-economy and governance structures, the major benefits of structural adjustment programs have probably been achieved. Further growth is likely to come from increased investment in the Samoan economy.

Samoa with plentiful, inexpensive labour; good regulatory, physical and telecommunications infrastructures; a business-friendly government and regional trade agreements[4] which give investors in Samoa preferential access to a larger regional market (Narsey 2006), would seem attractive to investors. However, the probability of significant external investment is limited because Samoa has no mineral resources, limited water, expensive energy and is a significant distance from markets. Samoa also faces continuing skill shortages because well-qualified Samoans can leave for higher incomes and social and professional opportunities in countries around the Pacific Rim[5] which impede growth.

One possible source of replacement human capital is well-qualified Samoans who have been born, or who have lived much of their life, abroad: a group referred to in this collection as the Next Generations. These potential returnees in the Samoan case are, indeed, the offspring of the first generation of emigrants; the 'one-and-a-half' generation, who left as children with their parents, and the overseas-born second and third generations. Many have vocational and professional skills, and access to social and professional networks, which could offset some of the losses incurred as their Samoa-born kin leave Samoa in search of the same things that led their own parents to leave the country; in short, a 'brain gain' to partially offset the 'brain drain'. Additionally, well-qualified Samoans who have accumulated assets and capital abroad could choose to invest these in Samoa. The investment of this human and social capital could contribute to Samoa's national growth and development, which raises the question of whether these people are likely to return to settle in Samoa.

Little is known about this question, however. Earlier studies of qualified migrants who had returned to Samoa to work, but had subsequently returned to New Zealand, identified a range of financial, professional and personal reasons behind their reluctance to remain permanently in Samoa (Macpherson

4 These are the PICTA and PACER agreements.

5 Their contributions do not cease completely: many continue to remit in either cash or kind to non-migrant kin but access to their professional and vocational skills cease.

1983, 1985). These studies provided insights into repatriation experiences of those who had lived abroad, but had some sampling limitations. They focused on people who had been born and raised in Samoa, had spent time living and training abroad, and had returned, often in mid-career, to work in Samoa, before returning to their metropolitan 'home-away-from home'. These studies did not consider those who had returned and remained in Samoa. Nor could they consider, at that time, the Next Generations of Samoans, who have been born or have lived most of their life abroad, and who represent a new source of potential human and social capital. Furthermore, their findings are dated, and much has changed in Samoa since their completion in the mid-1980s.

Most Samoan families are now transnational entities, with members in expatriate communities in the mainland US, Hawaii, New Zealand, Australia, Fiji, and smaller numbers working in other nations (Sutter 1995). As a consequence of increasing availability and falling real prices of both international air services and telecommunications, these transnational families are both more mobile and in closer contact. Families are now able to move personnel and resources between locations to take advantage of opportunities and to respond to collective needs in ways which led two authors to refer to them as 'transnational kin corporations' (Bertram and Watters 1985, 1986; Bertram 1993).

As an almost inevitable consequence of this increased movement and greater connectivity, many 'Samoan' families now have polyethnic memberships as Samoan culture offers no obstacle to ethnic intermarriage which has occurred since prehistorical time. These expanded family networks now connect Samoa to other potentially valuable sources of material and social capital in Pacific and Pacific Rim nations. Furthermore, increases in available levels of formal education, both in Samoa and abroad, ensure that most Samoan families now have access to more social capital. As a consequence of this increasing transnationalism, Samoan society has been extensively and directly influenced by globalization. New ideas, technologies and wealth which have flowed into Samoa from expatriate Samoans have found varying degrees of acceptance.[6] The Samoa to which the Next Generations might return is a very different Samoa from that to which earlier migrants returned. For various reasons, Samoa today may be as attractive and interesting to overseas-born Samoans as the metropolitan countries once were to their 'first-generation' migrant parents.

While the socio-cultural differences and distances between life in 'home' and 'expatriate' communities are no longer as pronounced as they once were,

6 Technologies which can increase agricultural productivity and new building technologies have found ready acceptance and high levels of uptake; new wealth has become readily incorporated in a range of contexts and has transformed traditional rank-wealth correlations; some new ideas including some religious ideologies have encountered significantly more resistance, however.

the economic differences are still considerable. While improved economic conditions have produced higher incomes in Samoa, incomes have also improved in New Zealand so a significant gap remains. Despite significant improvements in the availability of health and education services, and the institution of an old-age pension, there remain significant differences between social service provision levels in Samoa and the comprehensive health, education and social welfare systems in New Zealand.

This raises the question of whether Samoans who have spent all, or most, of their lives abroad might return to the country and contribute to its social and economic development. This question has obvious policy implications: knowledge of what led these people to return to Samoa and what keeps them there could be used to adjust policy settings to make their engagement more productive. One way of establishing this is to identify those who have made the move and to discuss their experiences. The obvious corollary is to speak to a second group who had returned to Samoa, but had subsequently returned to New Zealand, about their experiences and what had lead them to leave (or re-return). Our interest in this commission stems from our belief that this study might answer some of these questions for those involved in planning policies to meet Samoa's human capital needs.

Our Approach – Six Categories of Returnees

The second part of the chapter focuses on Samoans who were born or have been educated and spent most of their life abroad – in New Zealand in this case – and returned to Samoa as adults. The most important of these criteria was that people had been educated abroad, since formal education in New Zealand reflects modernist values and promotes a secular, meritocratic, individualism. This emphasis contrasts with some socio-cultural values which remain central to Samoan formal education, which promotes a religious, gerontocratic, hierarchical collectivism (Pereira 2006).

To improve our understanding and provide a more detailed categorization, a number of subsets, or 'clusters' of informants are identified, based on their primary motives for returning. Each of these is further divided into two subsets: those living in Samoa, and those who had spent time in Samoa but who had since returned to New Zealand. This was considered essential to examine both types of return and re-return experiences, which led some to remain while others chose to leave, or re-return. A concluding 'discussion' section attempts to identify some generalizable factors which shape these experiences and the factors which seem to underlie them.

During annual visits to Samoa between 1995 and 2007, we had talked to New Zealand-born and -educated Samoans who had returned to their parents' homeland. From these discussions, we distinguished clusters of returnees on the basis of their primary purpose for returning. These were; 'serving families', 'seeking culture', 'social idealists', 'professionals', 'entrepreneurs' and 'explorers'. Because this categorization utilizes *primary* purposes for return, it simplifies a complex reality: migrants often had multiple motives and primary motives could change over time, and/or depend on the context of the conversation.[7] The complex interactions between motives which drive people and shape their experiences are captured in Samoan novelist Albert Wendt's, *Sons for the Return Home* (Wendt 1973), which deals with the experiences and reactions of the family's sons to Samoa.

In planning this project, we assumed that we could locate a representative group of the *next generations* who could discuss and explain their experiences of life and work in their parents', and in some cases grandparents', homeland. This proved more difficult than anticipated: some who were normally resident were overseas and many whom we had met on earlier visits no longer lived in Samoa. This left us with a smaller group than we had intended to talk to and raised questions about why some had left, so we decided to add another group who had returned to New Zealand, to identify experiences which led them to leave Samoa.

The conversations which provided our empirical base occurred over a number of years, and include comments about experiences which arose in the context of more general discussions. Some material came from published material in which comments about experiences in Samoa were made in the context of more general discussions. Some data came from our relatives' accounts of peoples' comings and goings in the village. This account cannot, and does not, claim to cover the entire range of experiences, or argue that one set of these was more common than another on the basis of systematic sampling of various groups. Despite these limitations, the qualitative narratives of our informants do give us a sense of the range of experiences; help us to locate the causes of this variety and enable us to connect these with expectations which returnees had of Samoa, and motives for both their return and subsequent departures.

'Service to Family'

These returnees went to Samoa to support or care for relatives who had returned to Samoa. Since Samoan culture promotes loyalty and service, *tautua*, to family, it was no surprise to find people in Samoa to support family out of a sense of obligation. Samoan culture emphasizes, in particular, the duty of children to those who have raised them, those 'whose sweat they have eaten', and draws on biblical scripture[8] for additional sanction for this expectation.

7 Returnees who claimed to be enhancing appreciation of culture, a publicly acceptable motive, were later found to have been deported but were reluctant to reveal that less acceptable fact.

8 Exodus 20:12 is the verse most frequently quoted in support of this argument.

Some had joined parents who had returned to live in Samoa, often after retirement, and now needed assistance. A young man observed that he was happy to be able to 'repay' his parents by helping them run their small business in Samoa. A young single woman, who had gone to Samoa to look after her father, after her mother's sudden death, explained that she could not ignore her father's needs since he had gone originally to New Zealand so that she could have a 'better life' than she might have had in Samoa. A young couple, managing a plantation in Savaii, explained that they were in Samoa because a dying father had asked them to promise that they would return to take a chiefly title and live on the family's land as he had planned to. Another single woman noted that, while she was happy to help, she was in Samoa because, as a single woman, she was better able than her married siblings to relocate to care for her parents.

Several had returned to Samoa to look after grandparents to whom they had been close as children, and who wished to live out their lives in Samoa. They too explained moves in terms of meeting a responsibility to people who had looked after them during childhood. A young man explained that his parents were both working full-time while he was at primary school and his grand-mother went to New Zealand and looked after him and his siblings throughout their schooling. He recalled that she had left 'home' to live in another culture to look after her grandchildren, and commented on the irony that he in turn had left 'home' to live in another culture to look after her in old age.

Servants' experiences of Samoa also differed significantly. While all recognized their duties to their dependent relatives, their experiences were shaped by their attitudes to their return to Samoa: some had gone willingly and expected to gain personally and professionally from the experience. Some were enjoying the respect shown them by their Samoan kin for acknowledging and performing their duties to their parents. One young woman was surprised that Samoans considered her returning to care for her mother 'such a big deal'. After all, as she noted, her mother had told her that, 'all good Samoans looked after their parents', but speculated that maybe overseas-born Samoans were considered less likely to do this.

Others were thriving in the village and saw unanticipated benefits in living in Samoa for a period. A man living with his grandparents observed that he was also acquiring some knowledge and skills which could come in useful in his profession if and when he were to return to New Zealand. While servants' motives were more psychosocial than economic, some took work in Samoa to allow them to fund extended visits, and found that the workplace was a more 'social' space than those in which they had worked in New Zealand. An older man noted that there were not so many cliques within the office environment and that he seemed to enjoy more respect from younger people in Samoa than he did where he had worked in New Zealand.

Some servants were, however, less happy and determined to return to New Zealand when circumstances permitted. These people had interrupted careers and relationships to carry out their duties, but had few expectations of Samoa and confided that they would have preferred to take their parents back to New Zealand where they believed better services would have been available, and where they would have been able to continue their careers.

Some were finding communication with family and professionals difficult, and the lack of medical and social services to which they were accustomed frustrating, particularly where parents had high needs. Disagreements arose over the causes and treatment of their dependent relatives' illnesses: kin preferred to employ Samoan healers and medicine, where caregivers who were more familiar with western medicine preferred to use western diagnoses and pharmaceuticals. A middle-aged woman pointed out that her mother's diagnosis was clear and unambiguous and no Samoan medicine was going to treat it effectively. And yet, she noted, she was forever heading off relatives' attempts to introduce Samoan treatments, and then being accused of denying her mother treatment which could save her life.

Some returnees complained that local relatives, who usually made decisions collectively and after discussion, were critical of their more unilateral decision-making style and accused them of being *fia poto*, or of overestimating their wisdom. One of those noted that, from her nursing training and her experience as a caregiver, she knew what would work and saw no need to consult, particularly when there was a need to act promptly. Another took a commercial decision to deny further credit to customers at a family store in an attempt to recover outstanding debts, and was accused of 'acting like a *palagi*', or European, and of failing to understand the social and political dimensions of business in Samoa.

Others reported that local relatives suspected their motives in returning to Samoa late in life. One noted that his family seemed convinced that he had only returned to look after his father, who held a high title, to ensure that he could claim the family's chiefly title and the right to control its land. He noted the irony of this, particularly since, from what he had seen of his father's life in Samoa during his stay, he had no wish to claim either.

While only time will tell, most of this group seemed likely to return abroad when circumstances changed. Their time in Samoa was essentially a culturally-sanctioned interlude with a finite time frame. Those who were enjoying their time were more likely to reconsider remaining, but even this group believed that professional and personal opportunities were more readily available in the larger, overseas societies in which they had grown up.

Culture Seekers

A number of returnees were motivated by concerns about their incomplete knowledge of their parents' culture and limited fluency in their language. Since many Samoan migrants are proud of their culture and of their language, and argue that this knowledge is central to an 'authentic' Samoan identity (Anae 1997, 1998), this was not surprising. The circumstances which had led them to return to Samoa differed, but embodied common threads. A young woman noted that she was periodically made aware of this 'cultural deficit' because of her limited ability to communicate with her grandparents during their visits to New Zealand, and that they had encouraged her to go to Samoa with them and to learn the culture and improve her language competence. A young man explained that he had always envied his older brother who had spent his early years in Samoa and could 'speak Samoan properly' and 'interact more easily' with Samoan relatives and Samoans in general. He explained that he wanted to achieve that level of linguistic and social competence and to increase familiarity with his Samoan relations.

In other cases, the motive for increasing familiarity with culture was less connected with personal identity than with socio-political objectives. Migrant Samoans have reconstituted a range of Samoan institutions and practices in expatriate communities and have made access to these available to those New Zealand-born Samoans who have appropriate cultural and linguistic competences. Some 'culture seekers' sought to achieve these to take a more active role in 'tradition'. A single woman explained that since accepting one of her family's chiefly titles, *suafa matai*, she had begun to move in different socio-political circles and to manage family matters. She had realized that while her knowledge of Samoan language and culture were adequate in daily life, it was insufficient to allow her to exert much influence in these new areas. She hoped to extend this knowledge so that she could take a more influential leadership role in her family in New Zealand.

Others felt that the versions of language and culture to which they had been exposed in the expatriate Samoan community were somehow 'incomplete' or 'inauthentic'. A young woman explained that Samoan language and culture in New Zealand were often performed 'mechanically' and without much understanding of their origins and 'real' meanings. Those who sought to increase their understanding of language and culture believed that they could only fully understand the parent culture by becoming a part of it and experiencing it in its 'natural context'. The 'culture seekers' contended that understanding of the Samoan language, world view and lifestyle would allow them to, 'feel more completely and more authentically Samoan'.

The culture seekers' experiences were generally more uniformly positive than those of other returnees. This is not entirely surprising, because they had chosen to return and valued what they were able to learn by returning. They agreed that local relatives were generally impressed with their commitment to Samoan culture and their desire to learn more of their national language and society's world view. They were compared favourably with other Samoan visitors from abroad who made little or no effort to learn these things, were critical of what they found in Samoa and 'acted like *palagi*'.

Most noted that relatives willingly spent time and energy ensuring that their fluency improved, and they also helped explain cultural concepts and customary practices. Some returnees found the latter more significant when they saw them enacted and were able to participate in them. It was, as one young man noted, easier to remember who gets which part of a large pig after you have helped to dismember it and carried its parts to the various recipients. Time frames were shifted in the process: those who judged their competence by reference to their peers in New Zealand tended to see their progress as more rapid, while others who judged their new-found competence by reference to local peers tended to be a little more circumspect about their achievements.

There was, however, some variability in responses to what they had found. While most agreed that Samoan language in the hands of orators and pastors could be both powerful and moving, there was less agreement about the world view and lifestyle. Some found both generally acceptable, arguing that they had no right to criticize ideas and practices which had been part of the culture for a long time[9] and before they fully understood them. Others found that as they understood some ideas and practices more fully, they began to have reservations about them. Some were particularly concerned with gaps between public profession and personal practice in both religious and secular contexts. Some complained that when they sought to raise issues they found few willing to discuss these matters and more than a little resentment at their having presumed to suggest that Samoans could do things differently. A young man noted that it was probably harder for people like him to raise these matters because he was told that 'people from outside' *tagata mai fafo* couldn't understand these things, which effectively ended discussion and reminded them that 'being Samoan' is about accepting its central values.

Many New Zealand-born, who had grown up in an avowedly meritocratic society, found the gerontocratic emphasis in Samoan society difficult to accept. Both men and women who had grown up in an egalitarian society in which communication is open and often frank, found the omnipresence of chiefs and stress on formal communication somewhat daunting, and the exchange of formal pleasantries and gifts somewhat time-consuming. Some New Zealand-born women, who grew up in a less gendered society, found some of the local attitudes and behaviour toward women conflicted with both their own beliefs, and public professions about women. Some were prepared to accept that this was an individual, rather than a systemic, issue but noted that Samoan social norms prevented people from intervening in situations in which they felt intervention was called for. They too, however, held back from overt criticism even when they felt compromised.

9 Some noted that Samoan society had been forming for some 3,500 years and that it was not for them to attempt to change a system of belief and organization which had served Samoa so well.

The culture seekers' time frames and strategies varied. Most believed that total immersion would resolve the language fluency issues relatively quickly and that as their understanding grew, their ability to participate in and understand other areas of cultural practice would increase. Some had set aside between one and two years, while others believed that they could accomplish their objectives in less time. Since those whom we met started from very different levels of competence, some variability was always to be expected. Some had attached themselves to kin in villages in the belief that the 'authentic' language and culture were to be found in rural areas. Others attached less importance to location since as they noted, the language was the same in town as in the village and that people travelled to rural areas to take part in more esoteric rituals.

Social Idealists

Some returned to transform various aspects of the parents' homeland. Since some migrants had left Samoa because they were dissatisfied with various institutions in Samoa, it was not surprising that their children should have come to believe that these were in need of reform. Idealists seemed to accept that reform of society was a longer term project and planned to remain in Samoa longer than the other groups.

A common theme among these people was the shortcomings of the 'traditional' religious denominations: the Congregational Christian Church of Samoa (EFKS); the Methodist Church of Samoa and the Roman Catholic Church all of which are descendants of churches established early in the nineteenth century (Gilson 1970). Since these returnees had been exposed to more egalitarian styles of worship and more democratic styles of leadership and church organization in New Zealand, it was not surprising that these idealists focused on religious reform. Some idealists were committed to reform of the traditional denominations which had, in their opinions, strayed from the true way and had imposed intolerable financial burdens on their members. Most were also aware of the significant power which the 'mainline' churches controlled, by virtue of their close association with government, and the futility of open criticism of these bodies. The idealists sought instead to offer 'alternative' models of religious organization and worship.

One young evangelical pastor, whose father's family had, at some time in the past, been humiliated by a pastor's public comment about their lack of generosity to the church, had returned to Samoa and established an evangelical congregation which concentrated on teaching Christian scripture and made few financial demands on its members. He pointed out that this was intended to show Samoan people an alternative form of religious organization and worship. Another, who had taken note of her father's criticism that pastors in the mainline churches were the sole source of authoritative interpretation of scripture and that they 'ruled' the villages, had returned to set up a more egalitarian 'fellowship' in which scriptural truths were discovered collaboratively, and organizational decisions were taken by a group of elders. Several New Zealand-born Samoan missionaries in the LDS church saw their missions as ways of showing Samoans an alternative model of worship, and of organization, in which the heaviest and most public financial demands of the mainline churches were limited and mitigated. One young couple, aware of the limited roles offered to youth in mainline churches, had returned to set up a youth mission in which youth interests and needs were central theological and pastoral foci, and had set up various youth development projects.

Some New Zealand-born and -educated 'idealists' were engaged in the reform of secular institutions. The basis and commitment to reform per se in these cases varied significantly. Some went to Samoa to contribute in training programs as part of occupational and professional responsibilities. Some who worked, either directly or indirectly, for the New Zealand government have been involved, usually on a short term basis, in programmes aimed at the reform of governance and institutional strengthening. They felt that they could be effective advocates of good professional and technical practice and that they could contribute in this way to national development. Another young man, a public policy specialist, accepted a number of short-term consultancy roles in a public sector reform program which ran from the mid-1990s in Samoa. He undertook this position in the belief that Samoan civil servants would be more willing to accept advice from a Samoan, and to adopt 'best practice' because he could relate the benefits of reform in Samoan terms which would make more sense to them.

Several professional educators had set up consultancies in the educational field and were instrumental in the reform of both curriculum and teaching practice. They too believed that the Samoan educational system would benefit from reform and that the professional development necessary to achieve this would be more readily accepted and embodied in practice if it was delivered by Samoans. This typically involved them in a series of short stays in Samoa where they gathered data with which to develop curriculum and programs and ran workshops to train local counterparts and teachers.

It was not easy work, and some admitted that Samoan counterparts and trainees had not been as receptive as they had expected. This was put down to the fact that while upper level civil servants who engaged them understood and were committed to the project, it was difficult to obtain full support from those in the courses who resisted suggestions about transformation of work places, particularly where this resulted in a loss of power. Others attributed the difficulties to the fact that, because of the ways the contracts were structured, they were unable to establish satisfactory relationships and were seen to be 'privileged' and able to come and go without committing to Samoan society.

Another noted that, while resistance to reform was common, it seemed in this case to be connected with the fact that this advice was delivered by a person from 'outside', and underscored the fact that Samoans were inclined to make distinctions between 'local' and 'overseas' Samoans, or *tagata mai fafo,* and to be more or less critical of Samoans who acted as if they were superior and deigned to give advice to 'real' Samoans. This problem was, as one noted, compounded by the fact that some 'trainers' were younger and better qualified than those whom they were training which 'goes against the grain of Samoan culture'.

Some had found the task of 'reform' too much and had returned to New Zealand. Some were philosophical, saying that Samoans were entitled to change or not and that they would do so 'in their own time'. Others were more critical and pointed to the fact that well-placed Samoans in government and the public service were well served by existing arrangements and had no interest in change, and that they could, and did, both actively and passively resist change. This was often stated more in sadness than in anger: noting that by turning its back on 'innovative practices', Samoa was denying itself opportunities to be more 'efficient' and more 'progressive' and creating and distributing wealth more widely.

Professionals

In a small economy like Samoa, which is continually losing human capital to higher wage economies elsewhere in the Pacific Rim, opportunities exist for relatively rapid professional advancement. This may be attractive to young professionals who wish to gain more responsibility and experience earlier than they might within their profession abroad. The presumption that professional competence alone will make the return straightforward matter may be to under-estimate the significance of the social context of professional activity. The Wellington-born former attorney-general, the first New Zealand-born Samoan to head a government department (Heather-Latu 2003) was selected after only eight years' experience in the Crown Law Office in New Zealand. As she noted, her legal practice was only one dimension of the experience, '… the experience of working in Samoa has been life changing: the challenges, the voyages of discovery of family, of culture, of genealogy and the very slow dawning of understanding of Samoa. This has not been easy: these last few years been the best and worst of times' (Heather-Latu 2003: 211). Brenda Heather-Latu held the position for eight years before resigning in 2006 over government criticism of her office's treatment of electoral petitions, but remains practicing law in Samoa.

Lawyers, who in New Zealand might remain in relatively junior positions in larger law firms – carrying out mostly mundane work on salary, can become principals in their own companies in Samoa, which allows them to practice in a wider range of law, to generate higher incomes and to achieve higher public profiles. One informant who became involved in politics through her husband who was a cabinet minister, is now principal in her own firm in Apia and has become a prominent human and women's rights advocate and commentator on public matters.

The challenges of professional activity, and professional returnees' experiences, may come from different styles and practices of professional practice in Samoa and New Zealand. Some see standards of practice and conduct within Samoan organizations that would be unacceptable in New Zealand and have to decide how to react to that. One option is to accept the reality and to learn from it. Brenda Heather-Latu explained her experience in the following terms, '... above all, I have learnt from being in a place where everything is fundamentally, universally unashamedly run the Samoan way, truly Samoa in all things' (Heather-Latu 2003: 211). There may be, however, limits to this philosophy. Eventually, Heather-Latu was unable to accept attempts by the government to influence her office's professional practice, or its criticism when she delivered views that were unacceptable to Samoan government leaders, and resigned from the office.

This experience is not unique among professional returnees. A young engineer, concerned with the ways in which projects were influenced by local politics, noted that while he had to accept those practices at present, they would no longer be accepted when he became a project manager. He believed that this attitude was widespread in a generation of younger professionals who would take control of a number of government operations within a decade and would promote changes in Samoan organizational culture. A number of people in professional positions, including several academics at the National University of Samoa, accept that some frustration is inevitable but that changes in professional standards will occur in the longer term in Samoa and they plan to remain to promote these changes, accordingly.

Not all have been prepared to take such a long-term view, however. The former registrar of the National University of Samoa, formerly a senior police officer in New Zealand, also held that position for several years before resigning and completing an assessment of the Samoan police force before returning to New Zealand. Some returnees avoid these professional issues by working for regional organizations such as South Pacific Regional Environment Programme (SPREP), UNESCO and UNDP, which allows them to live in Samoa but to avoid political interference in their professional practice which sometimes produces problems for those working within the Samoan public services.

Entrepreneurs

Entrepreneurs have begun to return to their parents' homeland in the hope of taking advantage of opportunities which they believe are available in a small, modernizing economy. Indeed, the government has become increasingly active in encouraging such 'entrepreneurs' to return and has established a 'one-stop-shop' to provide rapid approval of applications and licenses.

A steadily increasing number of such returnees are found in Apia in businesses ranging from insurance brokeraging, restaurants and hospitality, to tourism and car hire. Some had been in Samoa for as long as 12 years, others as few as 12 months. A number had found dealing with local Samoans in business problematical because of cultural expectations of kin. They had found that they were expected to discount goods and services to kin, and to accept indefinite delays in payment as part of a reciprocal system of exchange which operates within kin groups. These entrepreneurs noted that these expectations can create problems for small businesses with limited cash flow and high start-up expenses to meet. They were aware of entrepreneurs who had given up because their small businesses, particularly trading stores, but also gyms and pool rooms, could not carry the levels of debt that cultural obligations entailed.

Some admitted that it was very difficult to turn requests from kin down since they were, at least early on, unable to distinguish between *bona fide* kin and others who claimed to be kin, because they had not grown up with all those who claimed to be kin. They pointed out that it was easier for people raised in Samoa to run businesses successfully, because they could distinguish between kin and non-kin and to use language and cultural knowledge in a way which minimized exposure to this type of risk.

This knowledge had influenced their decisions to go into businesses such as hospitality and tourism where clients were unlikely to be kin, where advance deposits were made, and payments were made in full immediately; and into other businesses whose clients were used to paying for goods and services. They noted that tourists and visiting business people found Samoa 'inexpensive' and were more than willing to pay accounts without complaint, while for Samoans with much lower incomes, the same charges were considered very expensive and were more likely to be contested. It was, for some, easier to deal with overseas visitors from New Zealand because, ironically, they had more in common and shared similar views on how to conduct business. Others had, with experience, found ways of securing payment in advance which were culturally acceptable to Samoan clients and saw the possibility of expansion into these areas.

This was confirmed by returnees who had established businesses but had since returned to New Zealand. The experiences which had led them to return differed. Some had realized that without certain language skills and cultural knowledge it was difficult to conduct business successfully in Samoa. Others had realized that they were at a disadvantage compared with local Samoan business people, who had social and professional networks which could be utilized for a range of ends, and which conferred significant advantage on their businesses. Several noted that in Samoa, local business people accepted that lower levels of return on capital were inevitable, because of the costs of 'cultural participation and compliance', and worked with somewhat different business models which factored in these 'costs'. Others noted that despite improvements in regulatory and physical infrastructure, there were still real supply-chain obstacles which prevented them from conducting business in ways to which they were accustomed. They discovered that the conduct of business was more complicated and that the high returns which they had anticipated from establishing business in Samoa were not being achieved.

Several had noted that it was easier to base the business in New Zealand, and to locate an agent in Samoa to deal with the complexities of Samoan culture in business on a daily basis. Indeed this evolving business model lay behind the recent ' Return to Roots' mission, promoted by the Pacific Island Trade and Investment Commission, which encouraged New Zealand–born entrepreneurs to visit 'home' islands to assess trade opportunities and locate local partners through whom to work (Nadkarni 2008). This model would make some of their social and financial capital available to Samoa, but would free them to deploy the rest in larger more profitable markets.

Explorers

Explorers typically visited their parents' homeland in search of the places, experiences and people at the heart of their parents' narratives. Kinship, or *aiga*, is central to Samoan social organization and to personal social identity, so it was no surprise that some explorers had gone to Samoa to establish, or re-establish, relationships with relatives. Some had sought out family whom they had met only briefly or had not met at all, but who had loomed large in parents' narratives throughout their lives. Grandparents and parents' siblings were the most commonly mentioned subjects of interest. Several returnees noted that these people had been represented as heroic figures in their parents' stories, and as people whose conduct they could well emulate, so that their visits to Samoa were an attempt to discover the person, or persons, behind such legendary stories before those involved died. These explorations, however, typically involved a limited period of time in Samoa. Most expected to accomplish the project in a year and to return to their 'homes' and to their careers.

Some who were enjoying the experience planned to extend their visit, or to return later to work for a longer period. The people they found were engaging; new experiences and mastery of new skills were extending their sense of identity; their local families moved to include them in family affairs, and Samoan kin seemed willing to give the returnees some latitude where they lacked linguistic or social skills. These people also seemed to be physically well, and able to handle less personal privacy than they had been accustomed to abroad.

Some planned to leave sooner than they had anticipated, because they were disappointed for a variety of reasons. This disappointment stemmed from an inability to communicate effectively with those whom they sought out. There was a sense of disappointment that what they found was not what they had expected. Some experienced physical discomfort with the village life; others with social constraints and financial demands imposed by their families; and from a sense that they were somehow 'different' and could never be a part of the community in the ways that they had envisaged. For many among these early leavers they had often found the climate and the lack of privacy more problematic than anticipated.

Discussion

Certain themes or threads were common in accounts of the experiences of all of the groups identified above. The first of these was the importance of the capacity to engage with the Samoan society they found themselves in on return. This involved an ability to engage with its religious and secular cultural values and practices. Both those returnees who had been able to engage and participate, and those who had not, agreed that this facility (or difficulty) to 'fit in' depended upon their different levels of cultural and linguistic competence. It turns out that 'engagement' very much reflects the various ways in which Samoa and Samoan culture had been represented to these informants during their childhood abroad. Studies of Samoan migrants suggested that there was considerable variability in the ways in which parents represented Samoan society to their children (Macpherson 1984), and the ways in which they exposed their children to Samoan practices in the 'home away from home' (Macpherson 1991).

Some who valued Samoan culture, and intended to return, had systematically exposed their children to the language and practice. Others, who migrated to avoid various elements of the culture, had promoted assimilation to New Zealand ways and customs and had deliberately excluded their children from Samoan culture. Between these two approaches was a less directive one in which parents had not pursued any cultural agenda, where exposure to Samoan language and culture was incidental and no clearly defined views of either 'Samoan' or '*palagi*' culture were systematically promoted. A number of autobiographical narratives (Fairbairn-Dunlop and Makisi 2003) revealed a similar variability. This range of approaches is reflected in a pattern of declining competence in overseas born Samoan in the language (Hunkin-Tuiletufuga 2001), which affects in turn the ability to understand and engage with cultural practices.

The ways in which Samoa had been represented to them by parents and grandparents influenced returnees' experiences of Samoa, but not always in the ways which one might have expected. Some of those who had been deliberately 'prepared' for engagement by their families were surprised with the reality which they found and noted that it was 'not quite what they had expected'. One such young man noted that his migrant father and mother had represented Samoan society as all that was good and pure, a sort of antithesis to the *palagi* culture by which they were surrounded. What he found was a very different reality which, he supposed, reflected the gap between the mythical Samoa which his parents had constructed to cope with their marginal social status in New Zealand and the rapidly changing Samoa which was increasingly exposed to intensifying global influences. Another woman, whose foster parents had presented Samoa as a stifling, calculating and conservative society, found a culture which she supposed reflected the gap between the reality and the Samoa that her foster parents had constructed to explain and justify the rejection of their language and culture of their homeland.

A similar gap emerged between the personal identities which individuals had assumed in New Zealand and those which were projected on to them by Samoans. Many had simply accepted that they were 'Samoans': their parents, teachers, and friends had all confirmed this (Anae 1997, 1998). When some arrived in Samoa, they found their claims to this identity contested by local Samoans, particularly when their views or conduct set them aside from local Samoans. While their Samoan descent was not questioned, their cultural authenticity was challenged, and they found themselves consigned to another category: *Samoa mai fafo*, Samoans from abroad, or *Samoa fananau i fafo*, people born outside, as opposed to *Samoa mao'i* or 'true' Samoans. This disparity between national identities affected people in different ways: some accepted that this was a matter of fact and that they would always be different in some respects from their island-born cousins. Others accepted that the differences were real and that it was for them to earn the right to be considered 'real Samoans'. Some others, however, were disappointed and confused by the denial of an identity that they had taken for granted and their assignment to a category of 'incomplete' Samoan.

Another influence on returnees' experiences was their family situation. Single returnees without dependants were least affected because they were free to move around and to make new friends without family constraints. They were able to use time and resources as they pleased and to associate with a range of people. Returnees with spouses found settlement more challenging. Their choices were constrained by spouses' needs and preferences which limited their ability to engage and participate in ways which were expected of them. One man noted that to be 'one of the boys' meant regular after-work drinking sessions and that his wife, also an overseas-born Samoan, discouraged this pointing out that she didn't want him to be like his Samoan colleagues. Those with non-Samoan spouses, felt additional pressure to explain and justify Samoan custom and practice, and to provide translations of much of what was happening, which, as one woman noted, was a necessary compromise but took the spontaneity out of much social interaction.

The presence of children also influenced settlement and feelings toward Samoa. As children approached primary school age some parents became concerned about whether Samoan education would limit their children's later education achievements. This concern was more pronounced for some as children approached secondary school age and they were led to consider whether the somewhat more restricted range of course choice would constrain their potential. While a number of those to whom we spoke had no children they shared similar views about the educational consequences of remaining in Samoa, although as they noted, they were reluctant to air these matters publicly.

Improved telecommunications have meant that returnees are no longer as separated from New Zealand as they might once have been. They have also had an interesting consequence on people's experiences of Samoa: those who were enjoying their experience pointed out that, as the costs of calls and texting had dropped dramatically, they could easily maintain links with family abroad and that had minimized any sense of 'isolation' from family. Those who were ambivalent about their experience explained that the ease with which they could maintain contact with friends and family abroad served to highlight their 'isolation' from personal networks and to contribute to 'homesickness'.

Similarly, the reduced airfares which have resulted from more open competition, have made it possible to travel between Samoa and New Zealand more frequently and more cheaply. Those who were settling into Samoan society found that they were finding more to do in Samoa and made fewer journeys to New Zealand; those who were uncomfortable in Samoa were making more trips 'home' and found these served to highlight the contrasts between their two 'lives' and were unsettling. Rather than helping to overcome 'homesickness', the visits tended to intensify it.

Conclusion

The group, referred to in this collection as the *next generations*, is an important one for Samoa. Its members possess a range of social, technical, commercial and professional skills, and capital, which could contribute significantly to Samoa's social, political and economic development. Furthermore, this stock of human (and social) capital is growing steadily, as a proportion of the skilled local Samoans who left the country for social, educational and economic opportunities in New Zealand and elsewhere- mainland US, Hawaii, Australia and Fiji – are now returning. This group's contribution will be most significant if its members choose to return to and settle in Samoa for longer periods of time. And, the signs appear positive that the Samoa these Next Generations might return to is not only very different from the country their parents left, but more attractive and interesting, with greater opportunities on offer. Their willingness to settle will be influenced both by their expectations of Samoa and Samoa's expectations of them. This chapter has sought to explore some of the dimensions in which adjustments may be required to facilitate their settlement and contribution to Samoa.

References

Ahlburg, D. (1995), 'Migration, remittances and the distribution of income: evidence from the Pacific', *Asian and Pacific Migration Review*, 4(2): 157–68.

Ahlburg, D.A. (1991), *Remittances and their Impact. A study of Tonga and Western Samoa*, Canberra: NCDS, RSPS, Australian National University.

Anae, M. (1997), 'Towards a New Zealand-born Samoan identity: some reflections on 'labels', *Pacific Health Dialog*, 4(2): 128–37.

Anae, M. (1998), 'Fofoa i vaoese: Identity Journeys of New Zealand-born Samoans', doctoral dissertation, Anthropology, University of Auckland.

Bertram, G. (1993), 'Sustainability, aid and material welfare in small South Pacific island economies, 1900–1990', *World Development*, 21(2): 247–58.

Bertram, G. and Watters, R.F. (1985), 'The MIRAB economy in Pacific microstates', *Pacific Viewpoint*, 27(3): 497–512.

Bertram, G. and Watters, R.F. (1986), 'The MIRAB process: some earlier analysis and context', *Pacific Viewpoint*, 27(1): 47–57.

Blumbaum, M. (1973), 'The Samoan Immigrant: Acculturation, Enculturation and the Child in School', doctoral dissertation, Education, University of Hawaii.

Fairbairn-Dunlop, P. and Makisi, G. (eds) (2003), *Making our Place. Growing up PI in New Zealand*, Palmerston North: Dunmore Press.

Fitzgerald, M. and Howard, A. (1990), 'Aspects of social organization in three Samoan communities', *Pacific Studies*, 14(1): 31–54.

Forster, J. (1954), *The Assimilation of Samoan Migrants in the Naval Housing Area, Pearl Harbor, Hawai'i*, Honolulu, Hawaii: University of Hawai'i, Manoa.

Franco, R. (1987), *Samoans in Hawaii: A Demographic Profile*, Honolulu, Hawaii: East West Population Institute.

Franco, R. (1990), 'Samoans in Hawaii: enclaves without entrepreneurship', in Connell, J. (ed.), *Migration and Development*, Canberra: National Centre for Development Studies, Australian National University, 170–81.

Gilson, R.P. (1970), *Samoa 1830–1900. The Politics of a Multi-cultural Community*, Melbourne: Oxford University Press.

Government of Samoa (1998), *Statement of Economic Strategy 1998–1999: Strengthening the Partnership*, Apia: Treasury Department.

Government of Samoa (2000), *Statement of Economic Strategy, 2000–2001: Partnership for a Prosperous Society*, Apia: Government of Samoa.

Government of Western Samoa (1996), *Statement of Economic Strategy, 1996–1998. The New Partnership*, Apia: Government of Western Samoa.

Heather-Latu, B. (2003), 'Daughter, and the return home', in Fairbairn-Dunlop, P. and Makisi, G. (eds), *Making our Place. Growing up PI in New Zealand*, Palmerston North: Dunmore Press, 205–13.

Hunkin-Tuiletufuga, G.L. (2001), 'Pasefika languages and Pasefika identities: contemporary and future challenges', in Macpherson, C., Spoonley, P. and Anae, M. (eds), *Tangata o te Moana Nui: The Evolving Identities of Pacific Peoples in Aotearoa/New Zealand*, Palmerston North, New Zealand: Dunmore Press.

Kotchek, L.R.D. (1982), 'Adaptive Strategies of an Invisible Ethnic Minority: The Samoan Population of Seattle', doctoral dissertation, Anthropology, University of Washington, Seattle.

Macpherson, C. (1978), 'The Polynesian migrant family: a Samoan case study', in Koopman-Boyden, P. (ed.), *Families in New Zealand Society*, Wellington: Methuen Ltd, 120–37.

Macpherson, C. (1983), 'The skills transfer debate: great promise or faint hope for Western Samoa?', *New Zealand Population Review*, 9(2): 47–77.

Macpherson, C. (1984), 'Samoan Ethnicity', in Spoonley, P., Macpherson, C., Pearson, D. and Sedgwick, C. (eds), *Tauiwi: Racism and Ethnicity in Aotearoa/New Zealand*, Palmerston North, New Zealand: Dunmore Press, 107–27.

Macpherson, C. (1985), 'Public and private perceptions of home: will the Samoans Return?', *Pacific Viewpoint*, 24(1): 242–62.

Macpherson, C. (1991), 'The changing contours of Samoan ethnicity', in Spoonley, P. Pearson, D. and Macpherson, C. (eds), *Nga Take. Ethnic Relations and Racism in Aotearoa/New Zealand*, Palmerston North, New Zealand: Dunmore Press, 67–86.

Ministry of Pacific Island Affairs (1999), *The Social and Economic Status of Pacific Peoples in New Zealand*, Wellington: Ministry of Pacific Island Affairs.

Nadkarni, D. (2008), 'Return to roots mission', *SPASIFIK*, January–February, 24: 32–4.

Narsey, W. (2006), 'PICTA, PACER and EPSs: weaknesses in current trade policies and alternative integration options', in Powles, M. (ed.), *Pacific Futures*, Canberra: Pandanus Press.

Pereira, J.A. (2006), *Aspects of Primary Education in Samoa: Exploring Student, Parent and Teacher Perspectives*, Dunedin: Otago University.

Pitt, D. and Macpherson, C. (1974), *Emerging Pluralism: The Samoan Community in New Zealand*, Auckland: Longman Paul.

Rolff, K. (1978), 'Fa'samoa: Tradition in Transition', PhD dissertation, Anthropology, UC Santa Barbara.

Shu, R. and Satele, A.S. (1977), *The Samoan Community in Southern California: Conditions and Needs*, Chicago: Asian American Mental Health Research Centre.

Sinclair, R. (1982), 'Samoans in Papua', in Crocombe, R.G. and Crocombe, M.T. (eds), *Polynesian Missions in Melanesia*, Suva: Institute of Pacific Studies, University of the South Pacific, 17–38.

Spoonley, P. and Macpherson, C. (2004), 'Transnational New Zealand: immigrants and cross-border connections and activities', in Spoonley, P., Macpherson, C. and Pearson, D. (eds), *Tangata, Tangata: The Changing Ethnic Contours of New Zealand*, Melbourne: Thomson Learning, Dunmore Press, 175–94.

Sutter, F.K. (1995), *The Samoans: A Global Family*, Honolulu: University of Hawaii Press.

Tuimaleai'ifano, M. (1990), *Samoans in Fiji: Migration, Identity and Communication*, Suva: Institute of Pacific Studies, USP.

Va'a, U.F. (1995), 'Fa'a Samoa: Continuity and Change. A Study of Samoan Migration in Australia', doctoral dissertation, Anthropology, Australian National University.

Va'a, U.F. (2001), *Saili Matagi. Samoan Migrants in Australia*, Suva/Apia: IPS, University of the South Pacific, National University of Samoa.

Wendt, A. (1973), *Sons for the Return Home*, Auckland: Longman Paul.

World Bank (1993), *Pacific Island Economies: toward efficient and sustainable growth*, Vol. 8, *Western Samoa*, Washington: World Bank.

Chapter 3

The Ambivalence of Return: Second-Generation Tongan Returnees

Helen Lee

Most Tongans who have settled overseas have maintained close contact with kin and remained involved in the nation of Tonga through a range of other connections, such as those with churches and schools. Most of these ties are maintained from a distance by remitting money and goods, making phone calls, sending letters and emails and so on. Many migrants, however, also make return visits to Tonga. Such visits may be to visit family or for particular occasions such as funerals and weddings, or for events in Tonga such as church conferences or the popular *Heilala Week* in July each year when Tongans enjoy beauty pageants, street parades and other celebrations around the time of the late king's birthday. These visits vary from a matter of days to months at a time; indeed, it can be difficult to clearly distinguish between 'visits' and return migration. Some migrant families maintain homes in Tonga, to which different family members return for varying periods. Some older Tongans, particularly women, move between Tonga and the diaspora to live with their dispersed adult children and grandchildren, sometimes for a year or more.

Another reason for return is the need to care for elderly relatives, a task that may be shared over time by different members of the family. However, compared to these various forms of temporary movement, far fewer Tongan migrants return with the intention of permanently resettling in Tonga. Some of those who do attempt to return soon move back overseas when they find themselves unable to maintain the standard of living they had as migrants. Others stay on in Tonga, and those who are not retirees tend to work in government departments, in the education and health sectors, or in their own businesses (Maron and Connell 2008; also see Connell, this volume, for a discussion of returned health professionals).

The children of Tongan migrants, who I will refer to as 'second-generation Tongans', typically maintain much weaker and more limited ties to Tonga than their parents and, as will be shown in this chapter, their attitudes and emotional ties to Tonga are more ambivalent (Lee 2006, 2007a). The number of either migrants or their children who return to live in Tonga is unknown, as no relevant data are collected by the Tongan government, but members of the second generation are less likely to 'return', and those who do, are more likely to experience problems of adjustment. Some are born in Tonga and migrate as young children, some visit on one or more occasions during their childhood and adolescence, but even those who have never been to Tonga speak of 'return' in the sense of returning to the 'homeland' of all Tongans.

This chapter explores the reasons for second-generation Tongans' return as well as the factors that prevent most migrant children from wanting to live in Tonga. The experience of return is described and a case study is presented of an unusual set of 'returnees': adolescents who move to Tonga for part of their secondary education, often unwillingly. Finally, the chapter considers the impact of recent changes in Tonga and whether these may encourage further return migration among the second generation. The chapter's focus is not on the impact of second-generation return migration on the homeland, which has been the focus of much of the broader literature on return migration, but on the factors that motivate or prevent such movements and on the experience of return.

The material presented in this chapter is drawn from data collected during a three-year research project investigating the transnational ties of second-generation Tongans in Australia. Data were collected during a period of field work from 2005 to 2007, and participants completed detailed questionnaires and, in most cases, gave in-depth interviews. The participants were 179 second-generation Tongans aged 18 to 30, located in five cities and one regional centre in Australia and 49 Tongans living in Vava'u and on Tongatapu in Tonga. Most of the participants in Tonga were aged between 18 and 30 and had grown up in Tonga, so they provide a comparison with those in Australia. Sixteen other participants in Tonga were 'returnees', 12 of whom were under 18 years of age. However, as data collection has only recently been completed, not all interview transcripts have been analysed as yet, so this account is based on data for 104 participants in Australia and 12 returnees attending high schools in Tonga. The chapter also draws on the author's previous research, since 1988, with Tongans both in Tonga and overseas; see Morton (1996) and Lee (2003).

Migrants' Children: Motivations for 'Return'

When Tongan migrants return to Tonga for holidays or special events they sometimes take their children with them, and as they get older the children sometimes choose to make these trips on their own, with friends or relatives close in age, or with organized groups such as in the case of church camps. Some who choose to move to Tonga for longer stays follow in the footsteps of their parents' generation who had returned: that is, they work in the government ministries or in schools or hospitals, or start their own businesses. Others spend time in Tonga as missionaries or working for non-government organizations. However, their motives for return can be rather different than that of earlier 'first-generation' migrant returnees, as will become clear in later discussion. This is particularly the case for those who see their move as a means to learn more about their heritage and Tongan culture, as part of their search for cultural identity.

There are other second-generation Tongans who move to Tonga because they are forced to do so: the adolescents sent to live with relatives and attend school, who are discussed later in this chapter, and those known as *tipoti* (deportees); (see Cassarino 2004, on returnees who are 'unprepared'). Taken together, these involuntary migrant groups might very well outnumber other second-generation returnees, although there are no relevant statistics available to ascertain this estimate's veracity.

Although illegal migrants of any age are deported when caught, the deportees of concern here are mostly young men, and occasionally women, who have grown up overseas and are deported after being convicted of crimes, often after a period of incarceration. They usually do not hold citizenship of the host nation, although in the US case some naturalized citizens have also been deported (*Tonga Now* 2007a). No deportees are included in the Tongan interviews discussed here – unfortunately some interviews arranged in Ha'apai could not go ahead due to a temporary closure of the domestic airline – however I had a number of conversations with deportees during my early research in Tonga in 1988–1989 and during return visits in 2005 and 2007. They typically find it extremely difficult to live in Tonga, as most speak little or no Tongan and are unfamiliar with *anga fakatonga* (the 'Tongan way'). Deportees tend to be shunned by the wider community because of their status as deportees and find it difficult to gain employment, yet most are eager to be accepted and recognized as Tongan.

One deportee, Sione Koloamatangi, formed the Ironman Ministry in Nuku'alofa in 2002 to support deportees and was able to obtain small amounts of funding. Koloamatangi stated at the presentation of one donation: 'Ironman Ministry is an inspired work of God that was birthed in my heart as I struggle through pain and hardship to adjust to my birth place and its culture' (Ironman Ministry 2003). A study of the deportees is being conducted by Joe Esser for his doctoral research at the University of Minnesota (still in progress at the time of writing) and is entitled 'Behind Bars, Beyond Borders: Transnational Deportees in the Kingdom of Tonga'. Esser reports that the Ironman Ministry moved to Vava'u but members have experienced significant problems because some Tongans in Vava'u were 'vehemently opposed' to their presence (personal communication, 6 December 2007). This attitude is also evident in a recent article on *Tonga Now,* a news website, entitled 'Deportees: a major threat to Tonga's communities' (*Tonga Now* 2007a).

Among the 104 research participants from Australia, there were many who had visited Tonga and some who had even lived there for periods during childhood and adolescence. Fourteen were born in Tonga and left at ages varying from infancy to 12; the one 12-year-old is properly referred to as a member of the '1.5 generation' but for the purposes of this chapter is being included within the second generation. Eleven of the participants were born overseas but went to Tonga as young children, some with their parents while others were sent or taken to live with relatives. Most of the participants moved back overseas at around the time they began school; however, in a few cases, they remained in Tonga for all of their school years and in some respects are more first than second generation, despite being born overseas. 21 participants from Australia had spent time in Tonga as teenagers to attend school, for periods ranging from six months to three years.

These various encounters with Tonga before adulthood do not have any predictable impact on the likelihood of return migration. However positively or negatively they view their experiences there is no clear corresponding pattern in participants' responses to the question 'would you like to live in Tonga?' Some expressed strong emotional attachments to relatives remaining in Tonga, with whom they lived or socialized as children or in their teenage years, but again this did not necessarily translate into a desire to relocate to Tonga. In fact, only three of the 104 reported that they had already spent time as adults living and working in Tonga, so the discussion below is based largely on the opinions of those overseas about why they would, or would not, want to live in Tonga.

Reasons for Return to Tonga

Very few participants said they would like to return to Tonga permanently, except perhaps towards the end of their working life or for retirement. In fact, only three said they definitely planned to move to Tonga before retirement, while ten said they would at least consider a permanent move – usually qualifying this by saying they would only do so if they were sure of employment with adequate pay. A further nine said they would like to retire there, although of course many of their parents also say they plan to retire to Tonga, but few actually do. Nineteen participants expressed interest in going for a year or so to work and experience the country, and to learn more about their heritage or 'help' in some way, but all were clear that they would want to return to live overseas afterwards. In all, then, 41 of the 104 participants (39 per cent) said they would consider return migration on a temporary or permanent basis. Many of the remaining participants were adamant that they would not even consider living there.

'Helping' Tonga

Not surprisingly, none of the participants saw Tonga as the place to make their fortune. They did not see Tonga in terms of business opportunities; rather, they tended to see their role as 'helping' Tonga and its people. Robyn (a pseudonym) is a 20-year-old woman from Canberra who was one of the few who was definite that she would like to move to Tonga. When asked why, she replied: 'Oh, just because I love to help people. And I know that they're less fortunate and I've got a lot I can give, you know.' Similarly, Jenny, a 25-year-old woman from Canberra, was interested in going to 'do something positive', which she described as: 'trying to change some old views that aren't helping them. You know, just to, little things like that, like I'd like to help the country.' Litia, who is 24 and lives in Adelaide, was born in Australia and went to Tonga with her family, including her Australian mother, at the age of seven and remained throughout her schooling before returning to Australia for her tertiary education. Sitting somewhere between the first and second generation, she felt drawn back to Tonga:

> I feel like there, there should be a time in my life that I should go back and contribute by trying to help the people in whatever occupation I chose to do ... on one side I want, yeah I wouldn't mind going living there again because I love it so much. At the same time knowing what I do know of how the living is, I'm like it's a big no-no for me. I'm like I wouldn't live there, but I would visit. So it's still, it's still in the cards, it depends, you know, how my occupation takes me. I mean, you know, if I go into teaching, then I'll probably end up going over there doing a year just to teach or something then come back.

Learning the Culture

Some members of the second generation who return to Tonga for long-term stays, but do not intend to remain there permanently, are seeking knowledge of the Tongan lifestyle and culture. One participant from Tonga was Sione, an 18-year-old male who had relocated from Australia the previous year to attend Tupou College, a Wesleyan day and boarding school for boys, known as Toloa. Despite having visited Tonga for several holidays, by the end of high school he could not speak much Tongan and only 'knew a little bit of the culture, but not much' so he decided to do an additional year of schooling in Tonga. He changed his mind but his parents pushed him to go so he enrolled in Toloa: 'just studying in Toloa because this is like the only school that can offer like a schooling life ... schooling in our culture'. However, he was finding it difficult: 'they treat us like dogs 'cause we're older boys you know. So we're supposed to be mature enough to handle it. So we do'. When he completed high school he planned to return overseas, where there was 'more opportunity'. Semisi, another student at Toloa who was only 14, chose to move from New Zealand at the age of 13 because he 'just wanted to see how my dad's other life was'.

Some participants in Australia also expressed a desire to spend time in Tonga to see what their parents' lives had been like before they migrated. Tonga is ever-present in the lives of most Tongans overseas, from parents' stories of their childhoods to the complex ongoing transnational ties, so it is not surprising that some in the second generation want to experience Tonga for themselves. Christine, a 21-year-old woman from Melbourne, wanted to live in Tonga for a while: 'just to see how lifestyle is over there, you know, to see what everyone is going through over there; that gives you a better understanding of the way things are over there compared to here'.

The Simple Life

Others spoke of returning to Tonga because they perceived life there as less complicated, less stressful and generally more relaxed, although most saw this as best suited to their later life and retirement. Rebekah, a 21-year-old woman from Adelaide, did not envisage returning to Tonga in the near future, saying: 'I just don't see the lifestyle that I kind of want there.' However, she added:

> Maybe when I'm older, I reckon it would be a great place to go. Especially … when like my kids all go off and get married and stuff, I reckon it'd be a good place to go and have your own life there. It'd be like once I'm old, I reckon it'd be cool to go over there and like set up a restaurant or, like, yeah, reggae sort of restaurant or something like that. You know on the beach or something. That'd be cool. But, I don't know.

Similarly, Anita, a 21-year-old woman from Brisbane, said: 'the only time I would ever go back to live in Tonga would be, you know once I have completed/achieved everything that I wanted here and when I retire … Yeah, so that's the only time I would wanna go back and live in Tonga – and when they have hot water too!'.

Rejecting Return

The wider literature on return migration includes various attempts to identify typologies of returnees (Cassarino 2004; Conway et al. 2005). However, a more fundamental distinction is that between individuals and families. Many individuals who migrate for work or study intend to return; some move back and forth at intervals and eventually return home permanently. Their primary kin ties are in the homeland and they may move relatively easily between locations. Individual migrants who marry overseas and start families there, and migrants who move overseas as a family group, tend to find return migration more difficult. The cost of relocating is higher, there are likely to be family members such as foreign spouses and foreign born children who are unfamiliar with the homeland, and there are more people to settle into schools and jobs. Tongans migrate as both individuals and in family groups, and although there are no relevant statistics, it does appear that individuals who do not marry overseas while studying or working are more likely to return to Tonga than those who have married overseas or taken their families with them from the outset.

Among the second generation it also seems more common for them to return as individuals, although the decision to return is complicated by the fact that their immediate family is likely to live overseas, so that moving to Tonga entails moving away from them, not returning to be with them as is the case for individual migrants. This is one reason they are more likely to return for finite periods, often a year or two, rather than with the intention of settling permanently.

Ambivalent Reception

There are other significant factors discouraging second-generation return, and whether they return to Tonga alone or with family members, the process of adjustment is likely to be even more difficult than for first-generation 'returnees' (see Maron and Connell 2008). It has been well documented that migrants who return 'home' often experience significant difficulties readjusting (Connell and King 1999; Potter et al. 2005; Thomas-Hope 1999). However, they do have the advantage of having grown up in the home country and are likely to have considerable knowledge of people and place. Members of the second generation who return seldom share this advantage and may struggle to adjust even if they have visited for brief periods beforehand (see Conway and Potter 2007 on Caribbean returnees). First generation returnees may be able to slip back fairly easily into a local mode of being that enables them to 'blend in' (if they choose to do so), whereas members of the second generation will inevitably seem 'different' to those in the home islands.

Initially, their experiences are often similar to second-generation Tongans who visit for holidays or special events (Lee 2003: 145–8). Both visitors and returnees typically receive warm welcomes from kin and are treated as special guests. However as time goes on, they are just as likely to encounter ambivalence or even outright hostility from other Tongans, who view them as outsiders and may call them *pālangi* ('white') or *pālangi loi* (someone trying to act 'white'), which is regarded as insulting. Maria is a 23-year-old woman from Australia who at the time of her interview was living in Tonga for six months on a church mission. As a self-described 'half-caste' she was usually referred to by Tongans as a *pālangi* and, as she added tactfully: 'sometimes you don't feel as welcomed as you would be if you were like in Australia or something, but, um, there is that home feeling', With a laugh she described her experience of return as 'challenging'.

Returnees can become the subject of gossip, particularly if their spoken Tongan is poor and they are unfamiliar with Tongan cultural protocols. Maria reported that she had been the subject of disapproving gossip because she spent time with male friends: 'Because like for us, it's just normal for us to just hang out with guys, you know, like it's ok for us to just mingle and hang, but Tongans just, they are just like, [whispers] "what are they doing with them?" You know, like what are those girls doing with all those boys?'

Both visitors and returnees can also find it difficult to cope with requests for goods and money from extended family members. Tongans who have lived overseas are often perceived as wealthy and the Tongan values of sharing and generosity are invoked by kin wishing to share that wealth. Susana, a 30-year-old female from Sydney, was born in Tonga and moved to Australia at the age of three. She visits Tonga each year and commented: 'when you go to Tonga you suddenly find these relatives that you never know [laughs] who ask things of you and you and you never bring your suitcase back, don't ever think of coming back with a suitcase of clothes because that doesn't happen.'

Economic Considerations

Some second-generation Tongans have vague ideas of moving to live in Tonga because their family has land they can use; however, most do not follow up on these plans. Others, whether or not they have access to land, are ambivalent about the idea of returning or are adamant they would not consider it unless they could manage financially. As Jack, a 27-year-old man from Canberra, commented:

> Maybe like if I had a lot of money and I could go there and do things and make a business and maybe help things out. Yeah, but not just as is because – you know, everyone wants to get out of Tonga for a reason. You know? Even people that are educated and that, can't stay in Tongan because of the friggin' system. So why would someone like us, me, who doesn't have anything want to go and live there?

Ambivalence about return is well expressed by Nina, a 26-year-old woman from Melbourne, who was unsure if she would return even if she was wealthy. When asked if she would like to live in Tonga she replied:

> No … oh, unless I'm rich. Not rich rich, but to have my own house, and have all the stuff I have here, there; but then I'd feel bad just knowing that maybe my next door neighbour doesn't have everything I have and then people would want to borrow stuff and then I don't know … I don't know if I could say yes.

Even those who idealize Tonga express a similar reluctance because its economic situation deters them. Anna, an 18-year-old woman from Adelaide, said of Tonga:

> I mean the simple life, that's a benefit. You know, where everything's less complicated, it's so less complicated. It's really easy going that way. You know, the environment is fantastic; you know, it's just so full of life and laughter and music and all their singing, really. But, umm, living there: not really. Not unless I was paid enough.

The most common reason for participants' reluctance to return to Tonga was their concern with living standards. David, a 19-year-old man from Melbourne, was sent back to Tonga at eight months of age to live with his grandparents until he was eight, but he does not want to return: 'it's too different, too much poverty'. Susana had returned to Tonga as an adult to teach English at Toloa for a year. She, too, does not want to return:

> Probably not, no ... if you think about it, like, you know, it's just thinking about how you're going to live, how you're going to survive because it's very hard to get money there, you know. Probably, but if you really think about and set a strategy of how you're going to make money, probably yeah, but at this moment in time as we're speaking, no, absolutely not.

Others were more succinct in explaining why they would not return to Tonga. Amy, an 18-year-old woman from Adelaide, simply said: 'It's a third world country, so I wouldn't wanna live there.'

Experiencing Return

The above perception of Tonga as a third world country was common among participants in Australia and many referred to its poverty and poor living conditions. For those who return for more than a brief visit, the economic realities of Tonga can be difficult to cope with, particularly in the context of frequent demands from extended kin as mentioned above. Most jobs are poorly paid relative to overseas wages, yet the cost of most consumer items is higher. Those who have to rely on purchasing all or most of their food, rather than accessing locally-produced food from relatives, find that it costs more than they are used to and the range of products available is far more limited. Basic utilities are very expensive and the cost of living is rising rapidly, with the country's budget stretched close to breaking point. Despite wage increases of up to eighty per cent for public servants late in 2005, many families still struggle to manage and the situation seems likely to worsen. For example, from the beginning of 2008 all government school fees increased by 56 per cent (*Tonga Now* 2007b) and similar price rises for goods and services can be anticipated.

To live comfortably in Tonga, therefore, one needs a high income or a great deal of assistance from relatives overseas. During a visit to Tonga in

July 2007, I spoke to a first-generation migrant who had recently returned with her husband after many years in the United States. Her husband had taken up a position working for the Tongan government so they returned to Tonga, bringing container loads of household goods to ensure they could set themselves up comfortably. They were wealthy enough to cope with the high cost of consumer goods and to travel overseas when they needed to stock up on items unavailable in Tonga. Other returnees have been able to similarly cushion themselves from the harsher realities of life in Tonga and their huge houses surrounded by high security fences symbolize their somewhat separate existence. However, their ostentatious demonstrations of success are likely to increase the demands of extended kin for a share in this wealth, a pattern King (2000) has identified as common to return migrants in many parts of the world.

Even wealthy returnees are likely to face difficulties coping in Tonga, from the poor quality of tap water, to the poor state of roads and the inadequate infrastructure. Many schools in Tonga are poorly equipped and can lack basic items such as textbooks, and the health system leaves much to be desired. Riots in Nuku'alofa, the capital, in November 2006, destroyed more than half the buildings, including the two large supermarkets and many other commercial buildings. This has further limited the range of products available and has generally made life in town more inconvenient as many shops have moved to makeshift locations on the outskirts of Nuku'alofa. A large new cinema burnt down in the riots, restaurants and cafes are generally expensive, and like other forms of entertainment, are limited compared to the large cities in which most migrants have lived, such as Los Angeles, Auckland and Sydney. Outside Nuku'alofa, in the rural villages and small towns throughout the archipelago, the range of goods and services is far more limited and living conditions can be very basic. Sela, a 20-year-old woman from Adelaide, captured the view of many in the second generation when she exclaimed: 'I'm glad I'm always the visitor and tourist and not someone who lives there!'

Youths 'Returned without Choice'

Young children, even babies, are sometimes sent or taken to Tonga by their parents and left with relatives for varying periods of time to enable their parents to work long hours overseas and not rely on formal childcare (Lee 2003: 86–90). Teenagers are also sent back, but usually for quite different reasons; most commonly because their parents are concerned about their behaviour and the friends with whom they are associating (see Menjivar 2002 for a similar situation in Guatamala and Miller Matthei and Smith 1998 for Belize). Another common reason is parents' concern about their children's Tongan language and cultural skills, as discussed below.

When teenagers are sent to Tonga to live for periods of time with relatives and to attend school they are sent either on their own or with siblings and are sometimes accompanied by an adult relative who then returns overseas. As in the cases of Sione and Semisi, above, some teenagers decide for themselves they want to go to Tonga for part of their schooling, but more often that decision is made without their input, leaving them unhappy and angry that they have been forced against their will. Occasionally they do not know they will be staying until the accompanying relative is about to leave and they are told they will not be returning overseas with them. Malita, a 23-year-old woman interviewed in Canberra, gave an example:

> … my cousin's friend in Sydney got sent back to Tonga for two years and his parents didn't tell him he was going. They actually said they were going for a holiday and that they had to come back and he had to stay, but he'd be coming home soon. And so he sort of, I think instead of straightening him out, it sort of, he hated it. He hated the whole thing because he didn't know he was going there and so, and so now, when people ask if he's from Tonga, he says no and doesn't want to have anything to do with that sort of … and I don't think it improved his behaviour at all.

Fifteen boys were interviewed in 2006 at Toloa and the outcomes for 12 will be discussed here, ranging in age from 12 to 18. They had moved to Tonga from Australia, New Zealand and the US, and only two had chosen to return. Ten therefore, were 'forced returnees.' Some had visited or even lived in Tonga earlier in their lives, but some had never been to Tonga before or had any connection to anyone in Tonga.

The most common reason for this cohort's return was their parents' concern with behaviour. Tim, a 16-year-old from the US, was sent back by his father for 'crimes and stuff' and Tavake, a 15-year-old from Australia, was also sent by his father, for 'doing a lot of bad things'. Similarly, Finau was 12 and sent from Australia by his mother 'to be a good boy'. In another case Josh, a 16-year-old from New Zealand, was sent to Tonga: 'for not doing good in my report card, my exams and too much wandering around and not staying home'. Previously, his brother had been returned for the same reasons.

Others are sent to attend school in Tonga because of the perception that the schools impart 'culture' and sometimes because of family traditions. For Ian, a 14-year-old from New Zealand, it was his grandfather who suggested he be returned because 'he came to school here and he learned many things in his life from this school. He wanted me to learn the same things'. Ian's father had not attended Toloa and regretted it, so he made his son attend although he did not want to. Similarly, Pita and his brother, aged 12 and 15 and also from New Zealand, were sent to attend school and learn the Tongan language, as was Jake, a 16-year-old from North America. Martin, a 20-year-old man interviewed in Canberra, reported that he, his brother and his sister were sent to Tonga against their wishes for six months to attend Beulah College, a Seventh Day Adventist school, 'to get to know the Tongan culture and the Tongan ways and see what being a Tongan is really all about'.

The practice of sending children to Tonga to attend school can seem somewhat paradoxical, given that one of the often cited motivations for migration is to access a better education system (Cowling 1990; Lee 2003: 16). However, many Tongans have deep sentimental attachments to their former schools in Tonga, as well as a fear of the influence of non-Tongan friends in schools overseas. These factors and parents' desire for their children to speak Tongan and understand Tongan cultural practices can outweigh their desire for 'better' education overseas. In addition, many Tongans perform poorly in schools in the host countries (Folauhola 2007) so parents sometimes send teenagers to Tongan schools in the hope their grades will improve.

Girls are also returned, many to attend the girls' equivalent to Toloa: Queen Sālote College, known as Kolisi Fefine. Like Toloa, it is run by the Wesleyan church. In some respects it is easier for them as the college is right in Nuku'alofa, whereas Toloa is some distance from town, out in the 'bush' (*uta*), but there are still strict restrictions on the girls and they are expected to work hard doing chores outside school hours. Tangi, a 24-year-old woman from Sydney, had been sent at the age of 12 to attend Queen Sālote College for a year, with her 13-year-old sister. Tangi explained: 'my mum and dad wanted me to learn about the culture and get to know my Tongan side'. She said she hated it at first and suffered 'culture shock' but came to enjoy the experience so much she did not want to return to Australia. She says the experience has helped her: 'I'm more proud to be Tongan'. Tangi was one of the few participants who was keen to live in Tonga, and she said: 'I want to go live in Tonga. Yep, I want to go live in Tonga so, I'm sort of looking for a connection, for a connection to make me go back and work there.' However, she qualified this, saying:

> My plan is to retire. I mean grow old in Tonga. Or find a man who has the same dreams as me and go back earlier and just teach or help out in the villages ... I see it as a long term thing and short term. Like, I wouldn't mind living there for a few years, coming back. The only problem is, I know you need money to survive over there. So I'd need money in the bank before I'd live there.

Young people sent to Tonga to attend school struggle with many of the difficulties other 'returnees' experience, but have the additional emotional difficulty of being there against their will. They often have to live with relatives they do not know before they arrive, either all the time if they are day students, or on weekends and during holidays if they are boarders. Life in boarding school is often a new and challenging experience, particularly as living conditions are very basic. They usually lose any freedom they had overseas to spend time outside school with friends and are expected to work hard at household chores outside school times, or at the school in the case of the boarders, who do gardening and other chores. Corporal punishment is still administered in Tongan schools, and while most of the students are accustomed to this at home, for those who have been educated overseas it can be a shock to find it occurring at school. As with the deportees, young Tongans sent to school in Tonga can be made to feel unwelcome and can be blamed for any trouble that arises. Just as the deportees were widely accused of instigating the 2006 riots, young returnees attending schools have been blamed for escalating violence in the conflicts between rival schools and for generally being a bad influence on youth raised in Tonga.

When asked about their overall experiences all expressed ambivalence, talking about 'ups and downs' but clearly trying, for the benefit of the older Tongan interviewer, to be as positive as they could. Most appeared to be trying to make the best of the situation they had found themselves in. Tavake said: 'I didn't expect coming here and staying on my own. Coming to school without no family. They're all overseas and it's just me staying here with some other family I don't barely even know, so yeah.' Yet when asked if being in Tonga had been a good or bad experience he replied: 'It's ok. It's a good thing 'cause I get to know family here.'

Other difficulties these participants described ranged from Tonga being 'stinky', hot and dirty, to the lack of shops, the poor roads, cold showers, mosquitoes and other pests, homesickness, and being mocked as '*pālangi*' and for not speaking Tongan. The benefits they identified included getting to know their kin and what their parents' lives had been like in Tonga, learning 'culture', respect and discipline. Not all felt they benefited from the experience; Haloti, an 18-year-old man now living in Sydney, was sent to Tonga for his third year of high school as a form of 'discipline' but admitted, laughing: 'I was a bit worse when I came back!'

Of the 10 boys sent to Toloa against their will, nine said they were eager to return overseas, where 'it's better', 'there are more opportunities', 'everything seems much easier' and, as one said of America, 'it's way better than Tonga!'. Even the two who had chosen to return to Tonga said they would leave again after their studies and only Pita (aged 12) said he was happy to stay on in Tonga. Some teenagers who are forced to return do not stay long at all because they cannot cope with the experience, and some of the boys interviewed mentioned others who had arrived at the school and were so homesick they were distressed and constantly attempting to run away from classes, and who usually went back overseas after only a few weeks in Tonga. When asked what advice he would give to other young people overseas, Tavake replied: 'I'd tell 'em to think twice before you go to school [here] or come here.'

Future Return

Some countries that have experienced significant emigration have established policies and programmes to encourage return migration, sometimes with support from a host nation. For example, Thomas-Hope (1999) describes the *Returning Resident Programme* established in Jamaica in 1993, which provides various incentives to return, and the *Return of Talent Programme* implemented from 1994, which facilitates skilled migrants' return. Diatta and Mbow (1999) show that Senegalese in France have been encouraged to return by an agreement between the French and Senegalese governments in 1975 which has supported various programmes – with varying levels of success – to help to finance 'reinsertion' and voluntary return to Senegal. They note that it is 'important to strengthen, in partnership with the countries concerned, a policy of assistance to return-migrant trainees' (Diatta and Mbow 1999: 254). It is unclear whether these programmes for Jamaicans and Senegalese are open directly to the second generation, although family members can accompany returning migrants so the second generation could indirectly benefit.

Within the Pacific, Niue is one of the most depopulated islands and in recent years has made various attempts to encourage migrants to return. Initiatives have included employment schemes and bank loans for renovating houses but they have been largely unsuccessful (Nosa forthcoming). Nosa argues that the main disincentives to return include the high cost of living, limited employment opportunities, pay disparities between Niuean and overseas wages, inadequacies in the health and education sectors and the difficult physical environment; all factors that also apply in Tonga.

To the best of my knowledge, no policies or programmes specifically aimed at encouraging return migration are in place either in Tonga or in any of the nations in which Tongans have settled. In 2006 the incoming Prime Minister, Dr Feleti Sevele, established a Department for Tongans Abroad within the Prime Minister's Office. The Department aims to strengthen ties to Tongans overseas, 'look after the interests and concerns of Tongans overseas' and 'attract investment from Tongans overseas, especially young, educated and financially well off Tongans in other countries' (Government Information Unit 2006a, online). Given Tonga's dire economic situation it is highly unlikely financial incentives are available to attract investors, except perhaps some tax benefits, and unless they are planning to invest in Tonga return migrants cannot expect any form of assistance. The major host nations – New Zealand, the US and Australia – would do well to look at programmes that have been put in place elsewhere to assist return migration as a way of encouraging a 'brain gain' to counteract the 'brain drain' that has so far predominated in Tongan migration.

This is an ideal time to consider such programmes, as Tonga moves slowly towards democratic government and works to rebuild the economic and social effects of the events of recent years, including the six-week public servants' strike of 2005 and the 2006 riots. Tonga is on the cusp of significant change, symbolized by the forthcoming rebuilding of Nuku'alofa, which could attract more Tongans overseas, both migrants and their children, to return to be part of the new nation building processes. Flights to and from Tonga have become considerably cheaper in recent years and with increasing access to the internet in Tonga, returnees would be able to retain close ties to family remaining overseas.

Enticing Tongans overseas to return, through financial incentives and programmes to support their settlement, would certainly be of benefit to Tonga and would fulfil the dreams of many second-generation Tongans to spend time in their parents' homeland. Just as short term labour migration schemes would be of enormous benefit to Tongans who seek opportunities to work overseas but do not want to migrate permanently, short term 'return migration' programmes could be beneficial for Tongans abroad. In both cases the nation of Tonga is likely to benefit, in the first instance through increased remittances and in the case of returnees through injections of skills and money into Tonga's economy and through the strengthening of ties between Tonga and the second generation in the diaspora. My research has clearly shown that the second-generation's level of remitting is far lower than their parents' generation (Lee 2007a, 2007b) but building their transnational ties in this way could conceivably raise the level of their financial support for Tonga.

In the absence of any schemes to encourage return migration, at present Tongans overseas wishing to return to Tonga must support themselves. If they bring skills with them they are more likely to find employment and if they can adjust to the challenges of life in Tonga they may find it a positive experience. Others, however, will be forced to switch their roles from remitters to remittance recipients, so they become, like many in Tonga, dependent on support from kin remaining overseas. In such cases return migration is of little or no benefit to either the returnee or Tonga.

Conclusion

The extent of Tongan 'return migration' depends on how the term is defined. If it includes those who live overseas while they study and then return to Tonga as a condition of their scholarship, people who leave Tonga temporarily to work overseas either through formal schemes or informal routes, and migrants who return for periods of months or years that are not regarded as 'holidays', then there is a considerable flow of return migration. If it is more narrowly defined as migrants who return with intentions of resettling permanently there is a very limited flow (Maron and Connell 2008). In the case of second-generation Tongans, apart from holidays and short-term stays for events such as church camps there is some movement to Tonga for school and work, but it is rare for such moves to be permanent.

Some migrants' 'second-generation' and '1.5-generation' children are keen to spend time in Tonga to 'help', learn the Tongan language, or be immersed in the Tongan lifestyle and 'culture', although most would only do so if they had sufficient income to live comfortably. Economic considerations are a key deterrent to return migration as are the related problems of poorly-funded health and education systems, inadequate infrastructure and a limited and expensive range of consumer goods. Social factors can also act as deterrents, particularly the rather ambivalent reception even short-term visitors can experience. Members of the second generation are acutely aware that they will always be regarded as outsiders in Tonga and while they may hold idealized images of Tonga as a laid-back holiday destination they seldom express a desire to actually live there.

Of course, there are those who do not have a choice but to return to Tonga, and the deportees and returned teenagers provide an interesting contrast to the young people who can weigh up all the pros and cons when deciding whether or not to spend time in Tonga. Second-generation return migration has only recently begun to receive serious consideration in the literature on migration and transnationalism, and within that work there has been very little discussion of involuntary movement. Further research is needed to understand the experience of people such as the deportees and adolescents who are sent to live in Tonga, yet may not speak the language, may be unfamiliar with cultural protocols and may not have a supportive kin network to assist their resettlement. There is also a need for research into their impact on Tongan society: are the highly negative views many Tongans have of both deportees and returned students warranted?

Deportees and adolescents sent to school in Tonga are likely to remain the most numerous of the second-generation returnees unless policies and programmes are introduced to actively encourage migrants' children to spend time living in Tonga. The establishment of the Department of Tongans Abroad in 2006 was a crucial first step towards this but Tonga's economic woes will make it impossible for the government to fund any schemes in the immediate future. The Australian, New Zealand and US governments are in a better position to do so, and would do well to consider the benefits that could be wrought from encouraging second-generation return migration, in relation to the much needed 'brain gain' and to the bigger picture of remittances as the mainstay of the Tongan economy.

Acknowledgements

This research is funded by an Australian Research Council Discovery Grant (2005–2007) administered through La Trobe University. The project's title is 'Pacific futures and second-generation transnationalism: a Tongan case study'. Most interviews were conducted by Dr Steve Francis, Meliame Fifita and Rebecca Tauali'i in Melbourne and Tonga, Meliame Fifita in Mildura, Christina Latu in Adelaide, Toakase Lavaka in Canberra, Maopa Latu in Sydney and Taniela Hoponoa in Brisbane. Sincere thanks to the many Tongans in Tonga and Australia who have been so generous with their time. Pseudonyms are used throughout this chapter when their interviews are quoted.

References

Cassarino, J.-P. (2004), 'Theorising return migration: the conceptual approach to return migrants revisited', *International Journal on Multicultural Studies*, 6(2): 253–79.

Connell, J. and King, R. (1999), 'Island migration in a changing world', in King, R. and Connell, J. (eds), *Small Worlds, Global Lives: Islands and Migration*, London: Pinter.

Conway, D. and Potter, R.B. (2007), 'Caribbean transnational return migrants as agents of change', *Blackwell Geography Compass* [online early October 2006], 1(1): 25–45.

Conway, D., Potter, R.B. and Phillips, J. (2005), 'The experience of return: Caribbean return migrants', in Potter, R.B., Conway, D. and Phillips, J. (eds), *The Experience of Return Migration: Caribbean Perspectives*, Aldershot and Burlington, VT: Ashgate, 1–25.

Cowling, W. (1990), 'Motivations for contemporary Tongan migration', in Herda, P., Terrell, J. and Gunson, N. (eds), *Tongan Culture and History*, Canberra: Research School of Pacific Studies, Australian National University.

Diatta, M. and Mbow, N. (1999), 'Releasing the development potential of return migration: the case of Senegal', *International Migration*, 37(1): 243–64.

Folauhola, C. (2007), 'Tongan ontology and Australian schooling', minor project, Master of Educational Studies, University of Adelaide.

Government Information Unit (2006), 'Establishment of Department for Tongans Abroad in the Prime Minister's Office', Press Release, 24 February, http://www.pmo.gov.to/artman/publish/article_90.shtml (accessed 26 February 2006).

Ironman Ministry (2003), Newsletter vol. 1, no. 1, hosted on Planet Tonga, http://www.planet-tonga.com/ironmanministry/Newsletters/IMM-Newsletter-July-2003/index.shtml (accessed 4 December 2007).

King, R. (2000), 'Generalizations from the history of return migration', in Ghosh, B. (ed.), *Return Migration: Journey of Hope or Despair?*, Geneva: International Organization for Migration, United Nations, 7–55.

Lee, H. (2003), *Tongans Overseas: Between Two Shores*, Honolulu: University of Hawai'i Press.

Lee, H. (2006), '"Tonga only wants our money": the children of Tongan migrants', in Firth, S. (ed), *Globalisation, Governance and the Pacific Islands*, Canberra: Australian National University E-Press.

Lee, H. (2007a), 'Transforming transnationalism: second generation Tongans overseas', *Asia Pacific Migration Journal*, 16(2): 157–78.

Lee, H. (2007b), 'Generational change: the children of Tongan migrants and their ties to the homeland', in Wood Ellem, E. (ed.), *Tonga and the Tongans: Heritage and Identity*, Melbourne: Tonga Research Association.

Maron, N. and Connell, J. (2008), 'Back to Nukunuku: employment, identity and return migration in Tonga', *Asia Pacific Viewpoint*, 49(2): 168–84.

Menjivar, C. (2002), 'Living in two worlds? Guatemalan-origin children in the United States and emerging transnationalism', *Journal of Ethnic and Migration Studies*, 28(3): 531–55.

Miller Matthei, L. and Smith, D. (1998), 'Belizean "Boyz 'n the 'Hood"? Garifuna labor migration and transnational identity', in Smith, M. and Guarnizo, L. (eds), *Transnationalism From Below*, New Brunswick, NJ: Transaction Publishers.

Morton, H. (1996), *Becoming Tongan: An Ethnography of Childhood*, Honolulu: University of Hawai'i Press.

Nosa, V. (forthcoming), 'The impact of transnationalism on Niue', in Lee, H. and Francis, S. (eds), *Migration and Transnationals in Pacific Island Perspectives*, Canberra: Australian National University E-Press.

Potter, R.B., Conway, D. and Phillips, J. (2005), *The Experience of Return: Caribbean Perspectives*, Aldershot and Burlington, VT: Ashgate.

Thomas-Hope, E. (1999), 'Return migration to Jamaica and its development potential', *International Migration*, 37(1): 183–205.

Tonga Now (2007a), 'Deportees: a major threat to Tonga's communities', *Tonga Now* online, http://www.tonga-now.to/Article.aspx?id=4883&Mode=1 (accessed 17 December 2007).

Tonga Now (2007b), 'Government school fees up by 56 percent'. *Tonga Now* online, http://www.tonga-now.to/Article.aspx?ID=4868&Mode=1 (accessed 17 December 2007).

Chapter 4
The Return of Japanese-Brazilian Next Generations: Their Post-1980s Experiences in Japan[1]

Eunice Akemi Ishikawa

In 2007, Brazil hosted the largest population of Japanese immigrants and its descendants – approximately 1.3 million people – of any country in the world. Most of these Japanese descendants' families are immersed in Brazilian culture and society. At the same time, many of the parents who emigrated attempt to maintain their Japanese culture – such as language, foods, beliefs and so on – at home and in their family lives. The children do not readily accept their parents' cultural beliefs, however, and in most cases they neither speak the Japanese language nor have they visited Japan.

The most recent phase of return migration from Brazil to Japan started at the end of the 1980s. Most of those migrants were Japanese-Brazilians (*nipo-brasileiros* in Portuguese, *nikkeijin* in Japanese), who came to Japan looking for better incomes. At present there are over 300,000 of these recent returnees living and working in Japan, mostly as unskilled manufacturing labourers. On the other hand, most of the young Japanese-Brazilians who migrated from Brazil to Japan after 1980 spoke only Portuguese, so now they are facing the same divergence at home between themselves and their children. The Japanese-

---

1 This chapter is a modified version of papers written and presented by the author in the following magazines and conferences: 'Migration movement from Brazil to Japan – the social adaptation of Japanese- Brazilians in Japan', *Chiiki Sougou Kenkyu*, 30–32, International University of Kagoshima, 1–9 (2003); 'The Ethnic Identity of Brazilian and Hawaiian Japanese-descendants', *IUK Journal of Intercultural Studies*, 5(2–3), 47–56 (2004); 'Japanese-Brazilians' life in Japan: the case of the second generation children', presented at the Conference, *Diasporic Homecomings: Ethnic Return Migrants in Comparative Perspective*, Center for Comparative Immigration Studies, University of California, San Diego, May 2005; 'The Japanese-descendants families in Brazil and Japan', presented at the 37th Congress of the International Institute of Sociology, Stockholm, Sweden, July 2005; 'Cultural and language barriers inside the families – the case of Japanese-Brazilian families in Japan', presented at XVI ISA World Congress of Sociology in Durban, South Africa, July 2006; 'Condições das Crianças e Jovens Brasileiros no Japão e suas Perspectivas', in *Cem anos de imigração japonesa: marcas na educação*, Editora USP (forthcoming 2008).

Brazilian children raised in Japan during this most recent period – 1980 to present – speak primarily Japanese and are more familiar with the Japanese culture than their parents who, in Japan, attempt to retain their Brazilian cultural ethnic identity and traditions. In many families, it is very common for the children to speak primarily Japanese, and for the parents to speak only Portuguese; thus, they often are forced to communicate with each other using both languages.

This chapter focuses on the experiences of young, second-generation and third-generation[2] Japanese-Brazilians who came to Japan as adults, as well as their third- and fourth-generation children raised in Japan. Together with a small number of unidentifiable fourth-generation children born to third generation parents, these post-1980s returnees can be collectively labelled as the Next Generations. The chapter, therefore, examines the experiences of this latest influx of Japanese-Brazilian parents and the adaptation of their children to Japanese society, including those who attended Japanese or Brazilian schools. Finally, it shows how their parents' stressful lives at home and at work influence their children's education. The author will also present cases of young Japanese-Brazilians who continue their studies in universities in Japan, who at present represent a small number, but who were able to continue their higher education endeavour despite all the problems and barriers they encountered in their lives, including those of living conditions and the Japanese school system. The method of research is based on oral interviews with Japanese-Brazilian families (adults and children) conducted in both Japan and Brazil.

History of Japanese Immigration to Brazil

Japanese immigration to Brazil started in 1908, just after Japanese immigrants were prohibited from entering the USA, and because Japan needed to continue migrant emigration, due to the poor condition of the country at that time. Most of these Japanese emigrants were from the western part of Japan, such as Kumamoto, Fukuoka, Okinawa, Hiroshima and Kagoshima, and they comprised three quarters of the total emigrant outflow. Many among this first wave of Japanese immigrants to Brazil went to the southeast and southern parts of the country, especially to the states of São Paulo and Paraná, where

2 Some of these latest post-1980s returnees might have had Japanese grandparents, or even great grandparents, who had emigrated from Japan to Brazil in the earlier decades of the twentieth century, so they are both second- and third-generation returnees by lineage. Such discrete family lineages and intra-generational immigration histories are not recorded in the data, however. Hence, use of the general term Next Generations is applied to this most recent cohort of post-1980s returning Japanese-Brazilians, although collectively they might best be referred to as the Next Generations.

they worked in the coffee plantations. Continuing until the early 1970s, Japanese emigration to Brazil occurred in three waves; the first wave between 1908–1923, the second between 1924–1941, and the third wave of post-World War II immigration from 1952 to the present. Changes in the immigration policy of the Japanese government directed these waves. There was no direct management by the government during the first term and many migration contractors were conducting business freely. Many immigrants left their home towns with the purpose of working and saving money for a short period in Brazil before returning. However, most of them did not return to Japan, and became part of the wider, diasporic Japanese society (Sociedade Brasileira de Cultura Japonesa 1992; Kikumura-Yano 1994).

From 1908, when the first group of Japanese immigrants went to Brazil, until the beginning of the 1970s, more than 240,000 Japanese entered Brazil. The current estimate of the Japanese descendant population living in Brazil, including those Japanese who originally emigrated from Japan (12 per cent) and the second (31 per cent), third (41 per cent), fourth (13 per cent) and succeeding generations, is 1.3 million, as mentioned at the outset of this chapter (Sociedade Brasileira de Cultura Japonesa 1988; Veja 2007).

To this day, Japanese-Brazilians have tenaciously clung to their Japanese cultural heritage and practices while living in Brazil. Currently, most Japanese-Brazilians are of the second, third and fourth generations, and most of them cannot speak Japanese. Also, many of them have never been to Japan. At home, parents and children speak primarily Portuguese. The few occasions where Japanese is spoken in the house are those where Japanese grand-parents live with the family, or where there are 'first generation' parents who emigrated from Japan to Brazil in their youth, but have maintained their Japanese customs as they have aged. However, most of the children speak Japanese as a second language, though it is often mixed with numerous Portuguese words as a form of children's pidgin (Portuguese-Japanese). On the other hand, outside the familial home, most Japanese-Brazilians have completely assimilated into Brazilian public society; in schools, social gatherings and professional places.

Cultural Adaptation and the Formation of Ethnic Identity

In this section, the cultural adaptation of the Japanese immigrants and their descendants living in Brazil is analyzed on the basis of the customs, sense of values, and beliefs. This examination also includes assessment of the influence of the new environment in which each individual grew up, and their education in families with Japanese customs utilizing the concept of 'cultural capital' (Bourdieu 1991). Japanese immigrants, although living in Brazil, uphold and revere their Japanese cultural heritage in terms of their self-identity. They transform aspects of their stocks of cultural capital into advantageous symbols in their societal adjustments. In their minds being 'Japanese' means being 'honest', 'hard working' and 'highly educated'. They believe that these attributes and identities differentiate themselves from the rest of the people in Brazil, and help them (the Japanese-Brazilians) to be socially superior to others (Ishikawa 2004).

'Cultural capital', as defined by Bourdieu (1991), is the way a person acts depending upon the resources they are given, using them to advantage or disadvantage in a cultural environment. For example, linguistic competence, knowledge, social class standing, educational level, and such are mentioned as types of cultural capital (Miyajima 1994). The cultural capital of these Japanese descendants, therefore, sustains their 'Japanese' identity, although they are living as members of Brazilian society. They are taught by their parents or grandparents about the 'virtuous Japanese' in the Japanese descendant community outside Japan, and regard it as a positive symbol (Ishikawa 2001). On the other hand, when they move to Japan from Brazil, the symbolic 'Japanese' identity is not recognized as advantageous. On the contrary, it is simply ignored by Japanese society, since people coming from abroad, although they are descendants of Japanese immigrants, are not accepted as part of the core Japanese society. In fact, they are considered to be foreigners by law and also by Japanese society, because of their differences in citizenship, language, customs, culture, and ways of thinking. The only similarities are physical appearances and the fact that they have Japanese blood. Unfortunately, these traits are not enough for Japanese-Brazilian returnees to be readily accepted by Japanese society. Accordingly, on return these Japanese descendants lose their confidence in being 'Japanese', and the identity of their home country (of birth) becomes stronger; that is, they feel that they are more 'Brazilian' than before, or are meant to feel 'foreign' and not at all Japanese.

Japanese-Brazilians Returning to Japan

In Japan there has always been a traditional social refusal to accept foreigners into the island's labour market and into society more broadly. The Japanese labour shortages of the 1980s and 1990s, however, attracted foreign labourers, mostly from Asian countries, who entered Japan as tourists and remained there after their visas expired. From 1990 onwards, and as a result of the revision of the Japanese Immigration Control and Refugee Law, illegal foreign immigrants and their employers were to be strictly penalized. On the other hand, this revised immigration law allowed Japanese descendants, up to the third generation, to live and work in Japan, even if they lacked professional skills. Such a visa for Japanese descendants enabled them to avoid the rigid Japanese Immigration controls that other foreigners and 'illegals' faced and allowed Japanese-Brazilians to live and work in Japan as unskilled manufacturing workers (Kajita 1994, 2005; Yamanaka 1996; Kawamura 1999; Ishikawa 2000, 2003a).

The volume of Japanese-Brazilians returning to Japan grew from the end of the 1980s to the beginning of the 1990s. This influx began in the late-1980s with the return of Japanese immigrants who had settled in Brazil, but who had retained their Japanese citizenship while living in Brazil with permanent visas, and later evolved to include Japanese descendants with Brazilian citizenship; mainly second- and third- generations. As a consequence of the partial reform of the Law of Entrance of Foreigners and Refugees in Japan, the Brazilian population in Japan, which was approximately 2,500 in 1987, increased to over 300,000 by 2007 (See Table 4.1).

The 1980s and 1990s coincided with a severe Brazilian economic recession followed by a lengthy and ever-so-slow rebound that encouraged many Japanese descendants to take advantage of their Japanese lineage to return to live and work in Japan (Reis 2001; Ishikawa 2003b). The second-generation descendants obtained visas qualifying them as 'spouse or child of Japanese'. The third generation or people married to a descendant of Japanese origin, received 'resident' visas. The fourth generation descendants do not qualify for 'resident' visas; although, there are exceptions for cases in which they are minors accompanying third generation parents. Thus, they aren't considered to be foreign labourers by Japanese Immigration Control, rather they are visitors who are coming to Japan to visit their relatives. However, most of them come to Japan with the specific purpose of working, albeit temporarily.

Japanese law does not permit foreigners lacking special skills to work in Japan (in contrast to professors, foreign language instructors, religious affiliates, business people, researchers, entertainers, athletes and such). However, this special nativist clause in the 1990 immigration law that only Japanese descendants can enter and work in Japan became a method for limiting the entrance of a great number of foreigners without professional skills into the Japanese labour market.

The current unemployment rate in Japan is approximately 3.8 per cent (Ministry of Internal Affairs and Communication 2007), but many Japanese do not want jobs as unskilled workers in small factories (automobile or computer parts assembly, food industry, etc.). As a result, there is a shortage of labourers in such jobs, thus providing opportunities for foreigners. However, Japanese society still doesn't accept foreigners as members of society, especially those who are visually (physically) different. So, one solution that many small factories use is to hire foreigners with Japanese features: the Japanese descendants.

From Sojourners to Permanent Residents

Almost 20 years after Japanese-Brazilians started to migrate to Japan at high rates, the number with 'permanent visas' had increased from 373 Japanese-Brazilians (0.2 per cent) in 1994 to 94,358 (30 per cent) by 2007. From this evidence, we can infer that many Japanese-Brazilian returnees are acquiring a permanent visa to prolong their stay in Japan. The advantage of this visa over the initial temporary permit many used to enter Japan is that it does not require renewal. Furthermore, it offers such 'foreigners' the right to live in Japan for an indeterminate period, and also contains no restrictions on the type of work they can perform. Further, foreigners are able to obtain bank loans for 30 to 35 years with this type of visa; enabling them to buy a house in Japan, for instance.

With regard to Japanese citizenship, many Japanese-Brazilians might be forgiven in assuming that their children who are born in Japan automatically receive Japanese citizenship. This is not the case, because the Japanese citizenship law is based upon the blood concept (*suis sanguinis*). For Brazilian children born in Japan to obtain Japanese citizenship, one of the parents must be a Japanese citizen, or have obtained it through the naturalization process. Consequently, a permanent visa offers the most stability and security for Japanese-Brazilian 'foreigners' who intend to live in Japan. On the other hand, a permanent visa does not give the 'foreign' holder any political rights, such as the rights to vote and to run for public office.

Table 4.1 Japanese-Brazilian population in Japan and visa types

Year*	1994	1999	2000	2001	2007
Brazilian population	159,619	224,299	254,394	265,962	316,967
Spouse/child of Japanese	95,139	97,330	101,623	97,262	67,472
Resident	59,280	117,469	137,649	142,082	148,528
Permanent	373	4,592	9,062	20,277	94,358

* In 1987, the Japanese-Brazilian population in Japan was only 2,500.

Source: Japan Immigration Association, *Statistics on the Foreigners Registered in Japan*, 1995 to 2008, with permission.

Life and Livelihoods in Japan

Most Japanese-Brazilians living in Japan have no contact with the distant relatives who migrated from Japan. In many cases, the families in Japan do not want their relatives 'back home' to know that they have immigrated to Brazil, and that they have returned to Japan for economic reasons to work in the unskilled labour market there. So, the move from Brazil to Japan is rarely influenced by family connections, or family networks. Most moves, on the contrary, are undertaken strictly for economic reasons. Most migrants work in factories as unskilled, manufacturing labourers, and they undertake temporary contract work, which does not provide them with job security or social and health insurance. They were welcomed by the Japanese government to fill the labour shortage in the kinds of jobs, known as 3D (dirty, dangerous and difficult), which are rejected by the better educated, Japanese youth, avoided because of their low status and 3D character, or just left unfilled.

Contract Work Conditions in Japan for Returnees

The Japanese-Brazilians, of which 48 per cent are between 20 and 40 years old, are mainly concentrated in Aichi, Shizuoka, Nagano and Gunma prefectures. Most are temporary unskilled workers in construction jobs and in car-parts, electrical appliance factories and such. The reason for this high concentration of Japanese descendants in these four prefectures is due to the concentration of automobile and electrical appliance industries in these particular areas. Most Japanese-Brazilians work under flexible contracts, in which their salaries are based upon the hours worked. Despite the fact that Japanese labour law mandates the inclusion of foreigners in welfare benefit systems, most of the agencies that employ Japanese-Brazilians do not provide such benefits; health and retirement insurance packages, for example. And, instead of the factories in which they work, most of them are employed by intermediary agencies (*empreiteiras*), which send these contract workers to factories for limited time periods. Japanese-Brazilians, therefore, are required to pay a certain percentage of their salaries as commission to these agencies. Under this type of contract, Japanese-Brazilians can be dismissed at any time, with no ensuing legal responsibilities, or compensation from, the factories that have employed them temporarily. In addition, the manufacturing companies only pay the agencies for the hours worked by the 'foreign' employees without paying for benefits, which makes their labour even cheaper. In short, Japanese-Brazilians turn out to be a cheap, temporary and disposable workforce for the aforementioned Japanese companies.

The author interviewed one Japanese employer, who was director of an automobile parts factory located in Shizuoka Prefecture. The director was very clear in affirming that even though a particular Japanese-Brazilian labourer had worked for the company for 18 years, and acknowledged the Brazilian's experience, the company had no intention of hiring the worker under a full-time contract. The reason provided by the director was that his company does not hire any foreigners on a direct contract. A follow-up question to reinforce this 'stubbornness' might go something like this: 'if this Japanese-Brazilian worker becomes a Japanese citizen via naturalization, would the company hire him, since he would then officially be a Japanese citizen?' Interview decorum meant it was not possible to ask this pointed question, but I firmly believe that the response would most probably have been some other 'justification' for not hiring such a naturalized Japanese citizen, since the advantage of hiring the now-available, 'foreign' labour force has kept the cost of unskilled labour as low as possible.

Housing Conditions in Japan

Japanese-Brazilians who are contracted to intermediary employment agencies normally live in housing offered by the same agencies. One nation-wide reason for this is the difficulty encountered by all foreigners when trying to rent housing in Japan. Indeed, many Japanese realtors absolutely refuse to rent to foreigners, citing the allegations that many of them do not obey the rules imposed by the housing owners, do not pay the rental fees, make a lot of noise, thus disturbing the neighbours, among other (perceived?) bad habits. Another factor that makes it difficult for foreigners to obtain a rental contract with realtors is the necessity of a 'guarantor' who is, preferably (according to realtor protocol) a Japanese citizen.

Considering the situation from the tenants' perspective, many Japanese-Brazilians actually prefer to live in the rental apartments offered by intermediary employment agencies because, although there is often little privacy, they only pay a monthly rent. To clarify, according to conventional Japanese rental agency rules, when a future tenant enters a rental agreement via a personal contract, it is necessary to pay, in addition to the first month's rent, an amount equal to approximately 5 to 6 times the monthly rent. The relatively large sum needed 'up-front' usually includes a two month deposit (*shikikin*), two months as gratitude to the owner (*reikin*) and one month for the realtor's fee (*tesuryo*). Therefore, if the monthly rent of an apartment is, for example, US$500, the tenant will have to pay an average of US$3,000 or more upon signing the rental contract, and only the deposit will be returned to the tenant at the contract's end. Commonly, however, the cost of cleaning and repair of damages to the property will be deducted, so that quite often the tenant receives little or none of their deposit.

There is also public housing available, offered by the local and prefectural governments. The rent is calculated based on the tenant's income, which normally varies from US$50 to US$300 per month. To apply for this public housing, a person must fulfil requirements such as living or working in the city where the property is located, living with family, having an income within a range stipulated by the government and, in many cases, being selected, since the demand for such subsidized housing is invariably and consistently higher than the supply.

Because of the working and living conditions detailed above, most Japanese-Brazilians live in close proximity to other 'Brazilian foreigners', making contact with Japanese citizens and Japanese society in general, more difficult. Many cities are already becoming known for their large number of Brazilian residents, sometimes positively, with an emphasis on *samba* schools during local festivals. However, it is more common that these 'Brazilian foreigners' residential concentrations are thought of in negative terms, because of increased conflict between native Japanese and Japanese-Brazilians; some of which has resulted in criminal behaviour. The majority of Japanese-Brazilians, therefore, are isolated from Japanese society with their daily and weekly routines being segregated and distinct; this isolation and enforced segregation coming about despite living in Japan for years. Additionally, many Japanese-Brazilians are still barely able to speak the Japanese language, regardless of their length of stay; such is the social exclusion they experience.

In a 2007 study sponsored by the City of Hamamatsu,[3] which houses the largest concentration of Japanese-Brazilians of any city in Japan – numbering approximately 20,000 residents in 2007 – 25 per cent of the informants answered that they have lived in Japan for more than 12 years, and a total of 57 per cent for more than six years. Related to their job contracts, 76 per cent answered that they are employed by intermediary agencies, and 32 per cent do not receive any health benefits. While 30 per cent do receive health benefits from the prefecture (*kokumin kenko hoken*), these benefits are only provided officially to self-employed or unemployed persons. For those who work in companies, there is a compulsory health and retirement plan (*shakai hoken*), however, it usually is not provided to foreigners by most companies. Regarding housing conditions, 47 per cent answered they live in housing provided by intermediary agencies, 22 per cent in apartments rented by themselves, 20 per cent in public housing, and 3.7 per cent in houses that they own. In this research, note the increase in the number of Japanese-Brazilians purchasing their own homes in Japan, normally with a 30 to 35 year loan. As a result, they tend to live among other Brazilian foreigners, in cities, or areas within cities, with several Brazilian restaurants, Brazilian shops, and Brazilian schools that provide them with a 'Brazilian environment' inside Japan, where they can live among co-nationals and speak Portuguese.

The greatest cultural obstacle that Japanese-Brazilians face in Japan, however, is the language barrier. Most do not speak Japanese fluently, and among those 'foreigners' who do speak the language, very few can read or write it competently. Problems caused by customs and heritage differences between the Brazilian and Japanese cultures are also common-place. The Japanese-Brazilians are culturally Brazilian and are not familiar with the customs and ways of thinking of Japanese people. And, while Japanese-Brazilians retained some Japanese cultural traditions and practices when living in Brazil, that doesn't mean their customs and customary practices mirrored those in Japanese culture 'back home'.

---

3 Report about 'The Conditions of Living and Work of South-American Foreigners in Hamamatsu' (2007), undertaken by the Department of International Affairs of Hamamatsu County Office and Shizuoka University of Art and Culture. In this research 2,582 questionnaires were distributed (in Portuguese and Spanish) and there were 1,253 respondents.

Japanese-Brazilian Children in Japan: Their Experiences

The Japanese-Brazilian population in Japan in 2007 was the third largest population of foreigners living in the country. The largest group is Chinese, with around 600,000, and the second largest is Korean, with approximately 590,000 (Japanese Immigration Association 2008). Currently (in 2007), the Japanese-Brazillian population in Japan totals 316,967 persons. Of this total, 68,291 (22 per cent) are under 19 years old and 37,444 (12 per cent) are under 10 years old. From these estimates, it is obvious that many of these children attend, or have attended, school in Japan. Compared to the Japanese-Brazilian adults living in Japan, their children learn the Japanese language much more easily and quickly. Many of these children go to public kindergartens, where they experience their first contact with other Japanese children, and most of them continue studying and learning the national language in Japanese schools (Ishikawa 2006).

However, even if they speak the Japanese language, most of them have problems following the regular disciplines. The main reason is that they do not have the background the other Japanese children have, and also because at home they don't have native-speaking parents to help them with their studies. Many of their parents can not speak Japanese and most of them can't read or write Japanese. Furthermore, because these parents are only in Japan for the purpose of earning money and returning to Brazil as soon as possible, they do not expect to live in Japan for the long term. To accomplish this rapid return, both parents in most families work all day (and, there are cases in which they work at night too, taking other temporary jobs – 'moonlighting'). It is scarcely surprising, therefore, that many Japanese-Brazilian, hard-working, or overworked parents spend very little time directly involved with their children's studies at home.

Another problem to be emphasized is that many of these 'neglected' children lose the desire to attend school, and in several cases these children may stop studying altogether. This specific problem appears to be closely correlated to the age at which they come to Japan (or if they were born in Japan). From research interviews with Brazilian families in Japan undertaken during the last 10 years, one finding is that the lower their age when they come to Japan, the smaller appears to be the educational problem. When such 'foreign' children come to Japan at junior high school ages, the adaptations to Japanese schools, Japanese language and the curriculum of the schools are much more difficult. When the children come to Japan at an age that coincides with the beginning of elementary school or younger, their adaptation occurs much more smoothly (Koga 1998). But that does not mean that these foreign Japanese-Brazilian children can keep pace with equivalent age groups of native Japanese children in their academic studies – many cannot. Yet another problem faces families in the home environment, where it is very common that the children speak mostly Japanese, and the parents speak only Portuguese, so they must communicate using both languages. Bilingualism helps families communicate, but it does not necessary translate into educational performance enhancement in Japanese schools where Portuguese is not a second language that is taught, except in rare instances.

There are a minority of cases in which Japanese-Brazilian children continue their studies in the colleges in Japan. The majority, however, do not even go on to complete high school. After they finish junior high school, many choose to work in factories like their parents, to save money with a view toward investing in some business in Brazil. This appears to be a common 'aspiration' among the Japanese-Brazilian communities working in Japan, but in reality many stay on to live in Japan for 10 years or more, with no specific plan or ability to put such dreams into practice. Re-return may be a common intention, but delay and postponement characterizes the 'myth' while the intention is only enacted by the few among the more elderly returnees, and their Japanese-born young, second- and third- generations do not see an eventual re-return to Brazil in their futures.

For these children it is very difficult to obtain Japanese white-collar jobs because most of them, although they have graduated from the Japanese school system, have poor Japanese language skills (writing and reading). Another option for the Japanese-Brazilian children's education in Japan is to enroll in Brazilian schools run by expatriates in Japan. The huge differences between the Brazilian and Japanese cultures and languages are a prominent reason behind this strategy. Parents, expecting to return after meeting their target earnings goals in Japan also plan for the time when their children return to Brazil. Accordingly, many temporary worker Japanese-Brazilian families opt to send their children to these schools as a 'stop-gap' measure and to make the best of their temporary sojourn.

Japanese-Brazilian Children's Education in Japan

Most Japanese-Brazilian children born in Japan, when they reach school age, enter Japanese schools and, in many cases, begin their studies at Japanese pre-schools. Many of them, in contrast to their parents, are more fluent in Japanese than Portuguese, but that does not necessarily mean that they can keep up with the pace and pressures of the curriculum at Japanese schools as most Japanese children do. Others attend Brazilian Schools, which have existed in Japan since 1995. In these Brazilian schools, all of the instruction is performed in the Portuguese language, and the curriculum is recognized by the Brazilian Ministry of Education. However, whether they attend Japanese or Brazilian schools, Japanese-Brazilian children face diverse problems of learning and cultural adaptation in Japan.

As recently as 2000, the Brazilian Ministry of Education had gone on record that there were no official Brazilian schools abroad. Then, later in 2000 the Brazilian government announced an exception to this general pronouncement in the case of Brazilian schools in Japan, and one school in Orlando, Florida (USA). The schools are private schools and 'foreign' or 'international' too. Some operate as branches of private educational systems in Brazil, and others are private schools started in Japan. By the end of 2007, there were 50 Brazilian schools in Japan recognized by the Brazilian government (*Ministério da Educação do Brasil*). They are located in those cities containing large concentrations of Japanese-Brazilians; which regionally comes to 15 schools in Shizuoka, 14 in Aichi, 12 in Nagano and nine in Gunma prefectures (International Press 2007). Some of the problems these Brazilian schools encounter are:

a. They are few in number and their geographical locations are inaccessible, and therefore many children are unable to get to them.
b. The schools are expensive, especially for the Japanese-Brazilian parents who generally work as unskilled labourers in Japan. The cost is approximately US$400 per month per child, plus the textbook fee. The average monthly income of Japanese-Brazilians in Japan is nearly US$2,500 for men and US$1,500 for women (data from interviews).
c. There are many students who enter and leave school during the academic year. This is because when or if the parents lose their jobs, they must take their children out of school.

On the other hand, Japanese public schools are free of charge, including text books, so that those Japanese-Brazilian families that are able to enroll their children in these schools only pay for the food and other occasional expenses not provided by these schools. Such expenses are miniscule when compared with the fees at Brazilian schools in Japan.

The main problem that Brazilian children face in Japan is a consequence of their parent's work and living conditions, that is, as temporary and unstable foreign workers who, in most cases, intend to return to Brazil. Because their parents work in factories and consequently have little to no time to take care of their children's studies, many children do not receive the attention necessary for a sound education. This factor is not solely a result of the parent's lack of interest however, as is often cited by Japanese school teachers, but because of their parent's work and living situations.

It is common to hear comments from teachers and principals of Japanese schools that Japanese-Brazilians are not concerned with their children's education, and that they simply use the schools as 'nurseries' or 'day-care' for their children during the working week. One of the reasons for this criticism, or 'misunderstanding' is that few Japanese-Brazilians attend parent-teacher meetings, or other events on the school campus, with language barriers being a prominent reason. Another reason is that although the children have difficulties with their studies, many parents do not appear to help them very much. What is not realized in this conventional criticism of foreign parents' neglect of their offspring's study habits, is that the 'lack of attention' is often due to the language barrier at home, and the inability of foreign-speaking parents to comprehend the children's homework and assignments.

On the other hand, there are some Japanese-Brazilian parents who, unfortunately, or because of the urgency of their own priorities, place a greater emphasis on their stressed and stressful workloads than on their children's education. This is because the objective of many Japanese-Brazilians is to return to Brazil as soon as possible (within three to five years being a common goal). Thus, many of them are concerned with earning the maximum amount of money in a minimum amount of time, and expect to re-return to Brazil, their adopted homeland and birthplace of their children. Their children's educational inconvenience is expected to be temporary, and duly fixed when they are all back home again (in Brazil), having 'made their fortune', or at least met the family's financial target.

In reality, and contrary to what is emphasized in the Japanese media, and/or among some (prejudiced?) Japanese educators, Japanese-Brazilian parents <u>are</u> very concerned with their children's education. Many of them do as much as they possibly can to provide the best education that they know of, and can afford in Japan, although the actual situation is far from satisfactory. Others hope to provide a better education for their children after they re-return to Brazil, as mentioned above.

At the same time, one issue to consider carefully is that many parents consider their children to be perfect in the Japanese language because they attend Japanese schools. This misconception occurs because many of the children learn to 'speak' in Japanese with fluency similar to Japanese children. However, simply speaking does not mean that the children have learned to read and write; the two skills, which are the minimum that they need to study in Japanese educational institutions.

The prospect for Japanese-Brazilian children in the future depends on where they settle. If they remain in Japan, as mentioned above, most of those who attend Japanese schools will probably work as unskilled labourers in the manufacturing industries (automobile, foods, construction, etc.), since, for white-collar jobs, there is an excess of Japanese workers, and the need for foreign workers is in the blue-collar job market.

Concerning children who attend the Brazilian schools in Japan, some of them are also working as unskilled labourers in manufacturing industries. In this case, they are in the same situation as their parents, that is, lacking basic Japanese language ability. If the children who attend the Japanese schools or Brazilian schools in Japan return to Brazil, they will have to face the Brazilian labour market with no specific skills or often, with no Portuguese language ability.

The adaptation of Japanese-Brazilian children in Japan is not at all easy; firstly, because of the language and customs, and secondly, because many of their Japanese-Brazilian parents intend to return to Brazil. However, the children have difficulty identifying with Brazilian society (especially those born in Japan), so they tend to choose Japan as their homeland. The problem is how Japanese society considers these foreigners in Japan. From a Japanese perspective, Japanese-Brazilians are still temporary foreign workers who will leave Japan after a finite period of time. That means Japanese-Brazilians are viewed as visitors, and this feeling is extended to their children. It is quite common to hear Japanese school teachers voice the opinion that the most important thing these foreign children can learn is how to communicate in Japanese just to enjoy their short time in Japan, and then when they return to their home country, they should resume studying diligently.

Young Japanese-Brazilian's Return Experiences in Japan

Most Japanese-Brazilians do not return to Brazil at the time that they predict, and among children who attend Japanese schools, it is a rare few who continue their education in Japan after junior high school. The majority enter the Japanese labour market as unskilled workers, like their parents. The main reason for this is that despite having graduated from a Japanese junior high school and being able to speak Japanese fluently, very few possess an adequately high skill in reading and writing the language. In Japan, there is no repetition of elementary and junior high school grades, consequently, many Brazilian children, some of whom, in extreme cases, cannot even read or write the Japanese language, receive junior high school diplomas.

In Maria's case,[4] currently 23 years old, she went to Japan when she was 12, after completing the sixth grade in Brazil. In Japan, she entered the first year of junior high school, which corresponds to the seventh grade in Brazil. She studied until the end of the second year at the junior high school, but because her father changed his job, the family had to move to a new city and Maria quit studying. Today, she speaks both Japanese and Portuguese fluently. However, she has difficulty reading and writing both languages. Maria is married, has a daughter and lives in Japan. Her husband, a Brazilian, works in a factory.

Another case is João, who went to Japan when he was 8 years old and finished junior high school there. Today, 22 years old and back in Brazil, he speaks both languages, but like Maria, has difficulties reading and writing both. João is single, works in an office, and his salary is 1.5 times the salary of an average, unskilled worker. (The average salary of an unskilled worker is considered to be very low in Brazil.)

With regard to children who attend Brazilian schools in Japan, since they follow the Brazilian curriculum, they also follow the criteria of the Brazilian law of education, that is, if the child does not fulfil the requirements of the current

4 The case studies presented in this chapter, like the examples of Maria, João, Pedro and Ana, are fictitious names, and are based on interviews conducted by the author from December 2006 to January 2007, in Japan and in Brazil.

grade, he will not progress to the succeeding grade. From research conducted in Brazilian schools by the author, because many Japanese-Brazilians remain in Japan as unskilled factory workers, rather than permanently returning to Brazil, many teenagers who have attended these Brazilian schools end up on the same path as their parents; prematurely entering the Japanese labour market as unskilled workers, especially because they have not mastered the Japanese language.

When they find a better job, it is normally as an interpreter working in an agency and/or factory, since their level of communication in Portuguese and Japanese is considered sufficient for the job, even though it is only at the level of conversational proficiency . Most of these interpreters work under the same temporary, insecure work contract conditions as other unskilled workers, and in many cases they simultaneously perform the same labour as other workers in the factories.

On the other hand, in cases in which the children have attended a Brazilian School in Japan and then returned to Brazil, they will be able to enter the grade following the grade they completed in Japan. This is possible because the Ministry of Education in Brazil recognizes many Brazilian schools in Japan. However, as mentioned before, many Japanese-Brazilians have no definite plans for returning to Brazil or, despite having tentative plans, very few successfully realize their intentions and re-return.

Currently, the biggest problem is related to the education and future of the children, who either went to Japan when they were small or were born in Japan, because Japanese society is extremely regimented with respect to educational level and social class. There exists a strict hierarchical division of social classes based upon educational attainment and professional standing, and it must be admitted that, at present, Japanese-Brazilians have come to be looked upon as one of the lowest classes in Japan.

In an earlier study, the author verified that some young Japanese-Brazilians have continued studying after graduation from high school (Ishikawa 2007). Eleven teenagers were interviewed, and despite several difficulties, as much due to their parents' work conditions as their adaptation to schools in Japan and learning the Japanese language, they successfully gained entry into universities or vocational schools in Japan. The common characteristics among these teenagers are that they received support and encouragement from their parents, both financially and emotionally, to continue their studies.

Pedro's situation is that of a boy who went to Japan when he was 8 years old, attended Japanese schools, and in 2006, was invited to enroll in a Japanese university. He passed the same examination taken by Japanese students of his age. Many universities offer a limited quantity of seats to foreign students under different requirements and exams, but in this case it should be emphasized that Pedro competed under the same conditions as Japanese students, and was accepted. On the other hand, despite speaking Portuguese with his family at home, Pedro prefers Japanese to Portuguese, since he has difficulties in writing and reading the Portuguese language.

Ana, who went to Japan when she was 14 years old, began her studies in Japan in the third grade of a junior high school, and proceeded on to high school. Ana says that when she was in high school and decided to apply at a university, many teachers tried to dissuade her, saying that her Japanese knowledge was insufficient for acceptance by a university. But Ana, with perseverance and hard work, was accepted by a university in Tokyo, much to the surprise of the teachers at the high school she attended.

Pedro's and Ana's cases are still exceptions among young Japanese-Brazilians, who are barely able to complete junior high school, much less advance to high school in Japan. It is expected that in the future the number of young Japanese-Brazilians with bachelor's degrees will increase and, consequently, bring better opportunities for work and life in Japan. And, such improvements in quality of life and work opportunities versus those of their parents are likely to become more common place among second-generation immigrants, despite their lowly class status at the moment.

Conclusions

For children and youngsters who are living or who have lived, in Japan for a long period, both returning to Japan or re-returning and going back to Brazil are very difficult and fraught with obstacles. The first concern is fluency with the Portuguese language, which many of these children have not mastered as a native language, especially those who attended Japanese schools. Those who have attended Brazilian Schools in Japan face problems of adaptation in the schools in Brazil, not only related to customs, but also with the class content. Second, is the re-adaptation (or adaptation) to the customs of Brazilian society. It is important to emphasize the problem of children and youngsters returning to Brazil with insufficient knowledge of both the Portuguese and Japanese languages, that is, they lack sufficient knowledge of one language upon which to base their reasoning. Frequently, children and youngsters speak both Portuguese and Japanese, but neither perfectly, that is, they cannot read or write either language perfectly (in accordance with their age and grade in Brazil), although they can verbally communicate with no problems in both languages.

For those who remain in Japan, the chances of upward social mobility, that is, joining the labour market in positions besides the unskilled labour force in factories, are still very remote because a strong educational background is essential. There are very few young Japanese-Brazilians in Japan who know enough of the Japanese language to compete with Japanese citizens for skilled jobs, as was mentioned earlier. However, because the Japanese educational system has no repetition until the end of junior high school, simply having graduated does not mean that the young Japanese-Brazilians are appropriately prepared for a position that requires, as a minimum, the ability to read and write Japanese.

Furthermore, for Japanese-Brazilian or any foreign children, even acquiring higher education does not guarantee a professional, or higher skilled, job in Japan. Another factor to consider is citizenship. Since the Japanese Law of Citizenship is based on the sanguineous relationship, Brazilians born in Japan do not automatically receive Japanese citizenship. Thus, 'lack of citizenship' is itself a sufficient disadvantage and serves as a cause for discrimination against Brazilian foreigners in the labour market. This occurs despite the fact that discrimination based upon citizenship, or lack thereof, is illegal under Japanese labour law.

One option for such young, more highly qualified Japanese-Brazilians is to obtain Japanese citizenship via naturalization. But, it is worth emphasizing again that to improve their chances of upward social mobility in Japan, a higher educational background, language fluency and citizenship must all be attained to afford them any chance of being considered for a highly-skilled position. Even then, the competition with native Japanese is likely to be unequal, because of ethnic discrimination, social distancing practices, and the perceived low class status of their immigrant parents' generation. For these first generations, many of whom do not have a high educational background in Japan, acquiring permanent visas or gaining naturalization, at least gives their children – the second and third – generations a better chance at social ascension within Japanese society. And, those children who are born in Japan already qualify to become Japanese citizens, and have better chances of obtaining a higher education than the earlier cohorts of young Japanese-Brazilians, who accompanied their parents to Japan. One of the results of the author's earlier (2006–2007) research on young Japanese-Brazilians, who had managed to enter universities in Japan, is that in most of the cases, the parents had a stable financial situation in Japan, even when they were employed as unskilled factory workers. For example, among the 11 youngsters interviewed, six lived in houses in Japan owned by their parents.

The most important thing to remind ourselves of is that these Japanese-Brazilians, who go to Japan with the dream of a better life in Brazil, or in Japan, are well aware of the difficult conditions they will face regarding work and life in Japan. Above all, those Japanese-Brazilian parents who have demonstrated concern about their children's education and future, have worked long hours, suffered set-backs and lay-offs at work, and have had difficulty adjusting to, and fitting in with Japanese society and the wide cultural dissonance between their own Brazilian culture and this new Asian one. One strategy many used to address this conundrum was to earn the greatest amount of money in the shortest possible amount of time with a prompt (re-)return to Brazil being part of their overall objective. Many stay, however, and continue to struggle on behalf of their children's future. Staying or returning, Japanese-Brazilian 'second-generation' parents, their children and even third- and fourth generations have not found the transnational lives and livelihoods they experience and the achievements they have accomplished easy or without considerable identity and adaptation problems. Theirs is a cautionary tale.

Acknowledgement

The author extends special thanks to Darrell Akimoto, for suggesting modifications to the original English version written by the author.

References

Bourdieu, P. and Passeron, J.-C. (1991), *Saiseisan* (trans. T. Miyajima), Tokyo: Fujihara Shoten [Bourdieu, P. and Passeron, J.-C. (1990), *Reproduction in Education, Society and Culture* (trans. R. Nice), London and Newbury Park, CA: Sage].

Época (2007), 'As 100 melhores empresas para trabalhar – 2007–2008', special edition.

Fujita, S. and O'Brian, D. (1991), *Japanese-American Ethnicity – The Perspective of Community*, Seattle, WA: University of Washington Press.

Hamamatsu City, Division of International Affairs (2007), 'Relatório sobre as Condições de Vida e Trabalho dos Residentes Estrangeiros Latino-Americanos em Hamamatsu, 2006' ('Living and Working Conditions of South-American Resident Foreigners in Hamamatsu, 2006').

International Press Newspaper (in Portuguese: edited in Japan) (2007), 12 May.

Ishikawa, E.A. (2000), 'Dekasegi taizaisha to jumin no aida de – nikkei nambeijin no chiiki shakai sanka' ('Between Dekasegi sojourn and residents – The participation of the Japanese descendants from South American in the Japanese Society') in Miyajima, T. (ed.), *Gaikokujin shimin to seiji sanka* (*Foreign Residents and their Political Participation*), Tokyo: Yushindo.

Ishikawa, E.A. (2001), 'Burajiru no nikkeijin shakai ni okeru konin kankei – nikkeijin shakai no gyoushuusei kyouka no senryaku to shite no kekkon' ('Marriages in the Brazilian Nikkei society – cohesion intensification of Nikkei society by marriage strategy'), *International University of Kagoshima Journal of Intercultural Studies*, 2(1): 37–47.

Ishikawa, E.A. (2003a), 'Migration movement from Brazil to Japan – the social adaptation of Japanese-Brazilians in Japan', *Regional Studies*, 30(2): 1–10.

Ishikawa, E.A. (2003b), 'Burajiru shutsuimin no genjo to imin seisaku no keisei katei' ('Brazilian emigration policy and its developing process'), in

Komai, H. and Koido, A. (eds), *International Comparison of Immigration Policies*, Vol. 1–3, Tokyo: Akashi Shoten, 245–82.

Ishikawa, E.A. (2004), 'The ethnic identity of Brazilian and Hawaiian-Japanese descendants', *International University of Kagoshima Journal of Intercultural Studies*, 5(2–3): 47–56.

Ishikawa, E.A. (2006), 'Kazoku ha kodomo no kyouiku ni dou kakawaruka' ('How families influence children's education'), in Hirota, T. (ed.), *Kosodate/shituke (Children's Education)*, Tokyo: Nihon Tosho Center, 290–303.

Ishikawa, E.A. (2007), 'Shingaku wo hatashita nikkei burajirujin no wakomono no gakkou keiken' ('Young Japanese-Brazilians who entered upper-level courses and their experiences in schools'), in Miyajima, T. (ed.), *Gaikokujin jidou/seito no shuugaku mondai to kazoku haikei to shuugaku shien nettowaaku no kenkyuu* (*Survey on the Relationship between Foreign Families and the Education of their Children in Japan and Social Network Assistance*), *Kagakukennkyuuhi*, Tokyo: Japanese Ministry of Education, Culture, Sport, Science and Technology) (2004–2006), 75–87.

Japan Immigration Association (1994 to 2008), *Statistics on the Foreigners Registered in Japan*.

Kajita, T. (1994), *Gaikokujin roudousha to nihon* (*Japan and the Foreign Worker*), Tokyo: NHK Books.

Kajita, T. (2005), *Kao no mienai teijuusha* (*Invisible Residents*), Nagoya: Nagoya University Press.

Kawamura, L. (1999), *Para Onde Vão os Brasileiros?*, Campinas: Editora da Unicamp.

Kikumura-Yano, A. (1994), *Amerika tairiku nikkeijin hyakkajiten* (*Encyclopedia of Japanese Descendants in the Americas*) (trans. M. Ohara), Tokyo: Akashi Shoten.

Koga, E.A.I. (1998), 'Rainichi nikkei burajirujin shitei no kyouiku to aidenthithi' ('The education and identity of Brazilian children in Japan'), *Nempou shakaigaku ronshuu* (*Annual Review of Sociology*), 11, Kantoh Sociological Society.

Ministério da Educação do Brasil (2007), *CNE Conselho Nacional de Educação*, http://portal.mec.gov.br.

Ministry of Internal Affairs and Communication (Japan) (2007), *Statistic Bureau*, http://www.stat.go.jp.

Miyajima, T. (1994), *Bunkateki saisensan no shakaigaku* (*The Sociology of the Social Reproduction*), Tokyo: Fujiwara Shoten.

Ninomiya, M. (1992), *Dekassegui*, São Paulo: Editora Estaçao Liberdade.

Reitz, J.G. (1980), *The Survival of Ethnic Groups*, Toronto: McGraw-Hill (trans. into Japanese by Kurata, W. and Yamanoto, T. (eds) (1994), *Canada taminzoku shakai no kouzou*, Kyoto: Kouyou Shobou).

Reiz, M.E.F. (2001), *Brasileiros no Japão*, São Paulo: Kaleidos-Primus.

Sociedade Brasileira de Cultura Japonesa (1988), *Burajiru ni okeru nikkei jinko chousa houkokusho* (*Survey of the Nikkei Population in Brazil*).

Sociedade Brasileira de Cultura Japonesa (1992), *Uma Epopéia Moderna – 80 Anos da Imigração Japonesa no Brasil.*

Veja (Brazilian magazine) (2007), 12 December.

Yamanaka, K. and Koga, E.A.I. (1996), 'Nikkei burajiurjin no nihon ryuunyuu no keizoku to ijuu no shakaika – idou shisutemuron wo tsukatte' ('The migratory movement of Japanese-Brazilians to Japan and their socialization'); *Ijukennkyu*, 33: 55–72.

Chapter 5
Bajan-Brit Second-Generation Return Migration: '... Where am I Supposed to be – in *Mid-air*?!'

Robert B. Potter and Joan Phillips

The focus of this chapter is on second-generation Barbadians, or 'Bajan-Brits', who have decided to 'return' to the birthplace of their parents, focusing in particular on their patterns of socio-economic and socio-cultural adjustment on their relocation to the Caribbean region. The research was based on fifty-two in-depth qualitative interviews conducted by the authors with second-generation return migrants to Barbados. At the outset, our analyses show that the migrants experience pronounced racial affirmation on their 'return' to Barbados – in other words, they benefit from a series of positive effects which are attendant on living in a predominantly black society following their move to the Caribbean. In addition, it is demonstrated how as a group, these youthful returnees are advantaged in the employment market, being perceived as professionally adept, and being characterized by an 'Anglo-Saxon work ethic'.

But the research was also instrumental in showing how in the wider social-cultural realm, these second-generation migrants experienced feelings of alienation and anomie. Thus, any problems that these second-generation migrants may have faced in the UK/USA/Canada in respect of their racialized identities, have generally been superseded by issues relating to their national identities on translation to the Caribbean. Thus, the research speaks of the liminal, hybrid and in-between racialized identities of the second-generation migrants. For example, most reported that they were acutely aware of the incidence of race typing and racism in their adopted societies, reacting not only to the existence of 'black-on-white' racism, but with a real measure of surprise to the occurrence of 'black-on-black' distinctions in society, involving the operation of colour-class gradations.

As a further instance, we consider the extreme problems that many of our informants reported with respect to the formation of new friendship patterns in the Caribbean. Friendship patterns appear to be particularly difficult to forge for female migrants who reported serious issues relating to both workplace and sexual competition. Thirdly, in what appeared to be very much grounded within the local Barbadian socio-cultural milieux, many of the migrants reported that they were frequently labelled as being 'mad'. This 'madness trope' can be interpreted as a means of 'othering' and 'fixing' the returnees in a manner that means that they do not have to be listened to, or their advice heeded.

The sentiments of one young migrant concerning her overall 'in-betweenness' as a second-generation migrant within Barbadian society can be regarded as indicative of the broad arguments presented in this chapter:

> I used to think sometimes, where am I supposed to be? Because people in England don't want you there, and here now, they don't want you here. And I used to think, where am I supposed to be – in *mid-air*?!

Background and Methodology

In research carried out in the eastern Caribbean in the late 1990s, Potter (2001a, 2001b, 2003a, 2003b, 2005a, 2005b) identified a cohort of 'Young Returning Migrants' to the Caribbean, most of whom could be described as 'Foreign-Born Returning Nationals'. These were second-generation Caribbeans, born in the United Kingdom, United States, Canada or elsewhere, of first-generation Caribbean immigrant parents, who for a variety of reasons had decided to return to the country, where one or both of their parents had been born.

This relatively new cohort of transnational return migrants had never been the focus of specific study, and although aware of the existence of such migrants, neither the Ministries of Foreign Affairs in the respective countries, nor the High Commissions in London had precise information as to the number of such migrants, their migration histories, or their employment and wider socio-economic characteristics (see Potter and Phillips 2002). As noted over twenty years ago, while nations collect precise statistics concerning 'aliens', the same seldom applies with respect to returning citizens (Rhoades 1979; Gmelch 1980).

An exploratory research project, the first of its kind in the Caribbean, was therefore carried out from 1999 to 2000, funded by the British Academy (Potter 2001a, 2001b, 2005a, 2005b). The aim was to provide the first analysis of this new cohort of migrants, its scale and socio-economic impacts. In particular, by means of open-ended semi-structured interviews, the project investigated the experiences and attitudes of such young returnees to the homeland of their parent(s). This preliminary research was carried out in Barbados and St Lucia and descriptive overviews of the findings of these pilot studies have been presented in Potter 2003a, 2005a, 2005b. Some 25 interviews were carried out in Barbados at the end of 1999, and 15 interviews in St Lucia at the beginning of 2000. Contacts were initially made via the various organizations assisting with the research,[1] and were snowballed thereafter.

The pilot research established a number of recurrent themes, which provided a context for the more detailed and informed investigation to follow. As already noted, the difficulty experienced by young female returnees in establishing friendship patterns with indigenous females was one of these recurrent themes. Moreover, these adjustment issues were very much related to the 'transnational', 'hybrid' and 'in-between' characters of the young returnees, who seemed to occupy liminal racial, social and cultural positions within Barbadian society (see Potter 2001a, 2001b; Potter and Phillips 2002, 2006a, 2006b).

Having identified the issues affecting this cohort of young returnees to the Caribbean in broad terms, an in-depth examination of the experiences and adjustments that were faced by them was undertaken. This research, under the title 'Social dynamics of foreign-born and young returning nationals to the Caribbean', was funded by The Leverhulme Trust, with the project extending from January 2002 to July 2006 (see Potter and Phillips 2002, 2006a, 2006b; Phillips and Potter 2003; Potter, Conway and Phillips 2005).

The principal target group for the study was *foreign-born returning nationals*, those who were born in the United Kingdom (or the United States etc) and who have decided to make Barbados their home. And, the present chapter deals with the experiences of these young returnees to Barbados. Such migrants can also be referred to as 'British Barbadians', and 'American Barbadians'. All those who have a Barbadian parent can claim nationality by descent. Another group are those who were born in the Caribbean, but who later traveled to the United Kingdom (or elsewhere) with their parents. If after ten years or more they return to live in the Caribbean, they also qualify as *Young Returning Nationals*. They are frequently individuals who return in their 30s with their families as part of an extended family return, with one important catalyst being their parents' retirement and relocation 'back home'. Generally, this is an expression of an ideology of return and often a mutual decision made by both the second generation and their parents. A third, but smaller group comprises those who are married to a Barbadian citizen. The project was primarily interested in the first group, but as noted below, a smaller number of informants from the other two categories comprising the spouses of the returning nationals and those born in the Caribbean who had returned were also interviewed.

The research also included detailed discussions with relevant politicians and policy-makers, searches of national newspaper archives and focus group discussions with members of the indigenous Barbadian public concerning their

1 These organizations included the Facilitation Unit for Returning Nations (FURN) (part of the Ministry of Foreign Affairs established to facilitate the return of nationals), plus local returning migrants' organizations.

attitudes towards returnees. However, the principal source of information was 51 in-depth, qualitative interviews conducted with foreign-born and young returning nationals. In all but one case, the informants were happy for the discussions to be recorded. The interviews were semi-structured in that they sought to cover all the major life domains associated with migration that had previously been identified in the pilot study. All of the interviews were fully transcribed and NUD-IST (Non-numerical Unstructured Data Indexing Searching and Theorizing) software used to assist with the qualitative analysis of the data. The interviews rendered invaluable insights concerning migration histories, family and socio-economic standing, motives for migrating, employment and academic background, and socio-cultural adjustments and experiences. As noted, the interviews were carried out between March and June 2002.

Turning to the sample, similar to the pilot study (Potter 2003a, 2005a, 2005b), the overwhelming majority of young returnees were females, some 38, against 13 males making up the sample. Nearly 63 per cent were foreign-born, with 29 out of the 32 having been born outside Barbados, in the United Kingdom. A total of 19 of the interviewees while having been born in Barbados, had been brought up overseas, the majority in the UK. For this reason, although the sample of 51 returnees contains one person born in the USA, one in Germany and one in another Caribbean territory, the blanket term 'Bajan-Brit' is used as a simple shorthand for the group in the main body of this account.

The average age of the young returnees was just over 33 years when they migrated, and they had been living in Barbados for an average of just over seven and a half years. Most had migrated to Barbados when they were in their 30s, with this age cohort accounting for roughly half the total sample. At the time of the interviews, the returnees had an average age of just over 40 years, with an age range extending from 21 to 69 years. With regard to race, 44 of the returnees categorized themselves as black and seven identified themselves as white. Virtually all the interviewees could be classified as middle-class based on their socio-economic status in the context of Barbados. With respect to their terminal level of education, for instance, virtually all had been to a further education college, and over 35 per cent had been to university. Most had jobs that required training and might be described as professional, with a fair representation of accountants, administrators and managers of various types. Notably, only two of the informants were unemployed at the time the interviews were conducted.

Turning to the wider family circumstances of the returnees, over half (26) said they were married and nine described themselves as having a partner. Just over 78 per cent had children, with one child being the average, a category that accounted for 18 of the interviewees. The majority, some 33 out of 51, had partners who were Barbadian locals. Turning to the parents of the young returnees, significantly, 77 out of the total of 102 were also living in

Barbados at the time the interviews were carried out, thereby emphasizing the importance of older retired returnees in promoting the return of their offspring as young/foreign-born returnees as part of an ideology of return. This was combined with the fact that many older retirees acted as informal care-providers for their grandchildren, and as a consequence their children in turn did not want to lose this informal extended network of support that had been available to them in the UK. Such circumstances were also reflected in the fact that 16 (31.37 per cent) of the returnees reported that they were living with their parents at the time they were interviewed.

Racial Affirmation and Relative Advantage

Racial Affirmation on Return

As might be expected, among the reasons for returning to the land of their parents, positive issues relating to race assumed importance among the Bajan-Brits. Several of our informants made the basic sentiment clear, stressing the affirmation of self within a numerically predominant black society and the feeling of liberation this engenders:

> I enjoy being here … umm … in comparison to London – in terms of blackness.

> It is easier here to be black. In England I grew up in a white area, went to a white college … .

> Here in Barbados … I never have to worry about being black.

One young Bajan-Brit emphasized the overall feeling of release afforded by no longer being part of an ethnic minority:

> I don't want to be too harsh, but I felt *normal.*[2] I felt like the *average person*.'

A good example was provided by Barbara;[3] she stressed just how uncomfortable she had felt on occasion during her school years in the United Kingdom. Barbara explained how in dressmaking lessons she had frequently been selected as the 'model', or 'clothes horse', with the teacher openly

---

2 Italics used in the narratives are for emphasis by the authors. Emboldening is for added emphasis on certain phrases.

3 Pseudonyms are used for the names of informants and workplaces to protect anonymity.

commenting in front of the class that if the pattern could be made to fit her, then it would fit anyone:

> When I arrived here and got off the plane – it was such a relief – to be among my own people ... to be with full-figured black women.

Other informants linked their satisfaction on return to their ability to immerse themselves in the black culture of their parents' birthplace. For instance, in one case, this was expressed in contrast to the 'extreme whiteness' of the individual's experience in growing up in the UK:

> Coming back to the birth place of my parents, that was something very important to me, and having the experience of living in a black culture. Because the part of England I was bought up with was *very white*. I was brought up in Kent, so there were no black people where I was growing up. And I wanted to experience black culture, whether it was in Barbados, somewhere else in the Caribbean, or in Africa.

The Black British and the Employment Market

Another positive reaction on the part of the returnees reflected the fact that a combination of a British accent and perceptions of a superior British work ethic seemed to have propelled individual returnees into advantageous professional positions quite frequently. Indeed, our investigations showed that many Bajan-Brits work in professional jobs, but without the benefit of the normal (appropriate) qualifications. For instance, while one returnee had a college certificate in child-care, she was managing a major hotel on the west coast of the island. The owner of the hotel is a white Englishman, and it seemed clear that she had been preferred for the job not least due to her British background.

It is very evident that in seeking and acquiring employment, many of the Bajan-Brit returnees appeared to be highly aware of the common perception held by Barbadians and international employers alike, of a superior 'British way of working'. It was also apparent that they were conscious that this perception is to their advantage in the job market. This appreciation was clearly illustrated by one of the interviewees who emphasized the attraction of her way of talking and her general approach to work:

> I was selling advertising. The clients loved me because I was different and yeah ... it was '... come see me and let me hear you talk'. So, I knew that I had something which was an asset to me. It was my work attitude. When everybody dashed off at 5pm, bang on 5pm, 6pm, 7pm it didn't bother me. I just loved what I was doing, so I quickly got promoted within the company, so therefore there was no need to go back. Professionally it is an advantage

> because of … I put it down to … this is my personal opinion … colonial days. A lot of Caribbean people still – although it is really changing, because it is a radical concept now – but they still kind of look up to the white person or people that are different to them. So for me, I got treated differently from a professional point of view when I was interacting with people, because I was slightly different and the difference was really the environment which I grew up in. I thought differently. I didn't think Bajan. My work attitude was just getting it done whatever it took. So for me, my whole approach was different and the business people tend to like that approach because it made them more financially viable. So professionally, I have done well, and fortunately, my organization is pro-training, pro-self development. They will give you a year, two years, three years off to do your degree … .I am then battling with a personal bias. Number one, professionally it seems that you are getting through easier than the local person. And the only reason why I did and many people like me did, is because the work attitude or work approach is different.

It is noticeable here, that the informant not only mentions that her accent and work attitude count in her favour in the workplace, but she also views these advantages as linking to the colonial era and the historical hegemony of whiteness within Barbadian society.

The narratives provided by several other young Bajan-Brits tended to reinforce this same view of the advantage bestowed in the job market by virtue of being British:

> Getting a job was extremely easy … you find a lot of offshore companies are drawn to people with a UK background or if you have lived and worked overseas, that is a plus … I see myself as British in terms of the work environment … I use it to my advantage in my career.

> I was lucky, I was just in the right place at the right time when I actually got the job at BA, and then as I said, the job came up with Virgin. I got that through my experience with having worked with BA in the UK, so I didn't find it difficult to get a job. I haven't actually gone out to look for a job, if you get what I mean. I haven't put myself out on the market you know. I haven't actually thought: let's see if I can try and do something else. I have gone straight back into the airline industry which was where I worked in the UK, so in that respect no, I didn't find it hard. But then I *haven't really tried*, I suppose.

> I came over, worked on a number of contracts in training … worked with Sandy Lane[4] and then applied for a job and got that as well. It worked

4 The top-ranking exclusive west coast hotel in Barbados.

> out quite well for me. I think that employers still employ Bajan-Brits over American Bajans. The only English thing, I am sure … not that jobs are easy to come by, but I think that employers once they hear that you are British, they think perhaps the work ethic is different in the UK, because you have to work hard in order to justify yourself. And you use your initiative a lot more …

Another defining characteristic that proved an advantage in the employment sector was the mode of dress adopted by the young returnees. In the following excerpt, a young Bajan-Brit tells how his perceived professionalism projected by his appearance helps in the job arena *vis-à-vis* the attire donned by local workers:

> I used to go to work in a shirt and tie 'cause the guys there wear polo shirts and jeans. So, I stood out. Cuff links, the whole shebang – dress hard!

Along much the same lines, another young Bajan-Brit recounts how his friends' discerning appearance and general conduct have facilitated their entry into managerial positions at relatively early junctures in their chosen careers:

> I have three English friends here, and they just came back. They supervise in (name of supermarket), but for the time that I knew they were living in Barbados they were never working. How did they get a job? Not because of accent, because of the way they conduct themselves. Guy's hair used to be long like an Afro; when he went to the job, all off, clean-cut, shirt and tie. You wouldn't think it is the same person. Only 17 years old, you know, and he supervising 45 years old people that were there in that company for years that could've get the job. Right now there are companies in Barbados that are hungry for young people coming in from overseas, 'cause they know that we have the training. All the guys that are foreign in the company they always get through quicker, 'cause Barbadians sit down and watch you – like this. That is all they do, they just watch. And, you can always tell where they are, 'cause you could watch them.

As the above excerpts indicate, young Bajan-Brits recognize the distinct advantages that their English work ethic and other defining characteristics bestow. Thus, it seems that the young Bajan-Brit returnees use these advantages to successfully negotiate the employment sector.

Hybridity, In-Betweenness and Alienation in the Socio-Cultural Realm

But a principal finding of our research was that away from the workplace, our second-generation return migrants generally reported that they felt outside mainstream Barbadian life on a day-to-day basis. These feelings of relative 'disconnect' varied from those of being acutely aware of a hybrid identity and living between two worlds to feelings of downright isolation and effective anomie. And, they affected many areas of the young returnees' lives. Experiences ranged widely, with respect to difficulties of making friendship, unexpected experiences in the arena of race and race-typing and in respect of the returnees frequently being accused of being 'mad' by Barbadian nationals (Potter and Phillips 2006a, 2006b, 2008). All of these were specific domains in which a more general sentiment was expressed; specifically, that while issues of their racial identity had been salient in the UK, in Barbados these difficulties had been replaced by issues of national identity, 'othering' and anomie.

Difficulties in Making Friends

Difficulties in making friendships with local Barbadians were mentioned by the majority of young return migrants to Barbados in both the pilot and the main surveys. Making friends in Barbados was recurrently seen as very difficult, with a measure of relative alienation from mainstream Barbadian society being implied:

> In Barbados, friendship patterns are fixed.

> It's the clique mentality of Barbadians. Nobody ever phones you up and asks you out.

Notably, an important causal role was ascribed to friendship patterns in Barbados having been established via a British-style, selective pattern of secondary school admission and attendance. Barbados still has a common entrance examination at age 11, a very clear and strong pecking-order of secondary schools continues to exist, and it is apparent that these act as primary agents of socialization. As one young returning national explained:

> In Barbados school friends and who you know are very important. It is tough if you do not know anyone.

The same sentiment was expressed from a different perspective by a returnee born in the USA who had moved to the island with her Bajan-born husband:

> My husband has a lot of friends from High School and they helped a lot when we came down here. They made for a smooth network.

The importance of other returning nationals and expatriates was stressed time and time again, and it was also implied that social relations are easier in the workplace than in the social realm:

> I would not see my work colleagues socially. There are only four people in Barbados I would phone. They are all English of Bajan parents and have lived here for less than four years.

> My friends are all from outside Barbados.

Naturally, having strong family ties, marriage and targeting friends were all considered to be factors that make matters easier, but virtually none of the informants claimed to have female Bajan friends, regardless of how long they had lived in Barbados:

> I was very lonely at first. All my friends came from the UK. But I married here.

> I have not found friendship patterns too big a problem, but I have strong family support. Plus, my best friend is another UK-based returning national who left almost exactly the same time as me.

A repeated observation was that Bajans go to work and then go straight home without socializing. Several young returnees said that they had suggested to their colleagues going for a drink after work on a Friday, but that this had been greeted by a measure of incredulity. Barbados is traditionally seen as a conservative and quasi-secret society, in which people keep their private lives and problems to themselves (Lowenthal 1972). Bajan pride has frequently been commented upon (Springer 1976; Lowenthal 1972: 10, 279–80). Lowenthal (1972) cites O.R. Marshall as observing that 'Jamaica has a difficulty for every solution, Trinidad a solution for every difficulty: Barbados has no difficulties!' An associated contention among some of the informants was that Barbadians are 'standoffish' in their interactions with outsiders.

The principal difficulty expressed by the majority of young female returning nationals was in making indigenous female friends. Several informants were unequivocal in attributing this to rivalries; sexual rivalry with respect to finding a male partner, as well as economic rivalry in the job market:

> It is difficult to make female friends. Bajan women think you are going to take *their men* and *their jobs*.

The informant who had lived the longest in Barbados, for over fifteen years, stated:

> I now have a *few* Bajan friends. Many think you are *taking things away from them*.

Some informants explained the problems experienced in making female friends in terms of gender vistas, plus issues of social standing:

> There is some bigotry. Womens' vistas are more limited here ... being secretaries, having children, being mothers. They are suspicious of outsiders. I expect they think we see ourselves as better than them. And in the workplace they are worried about their jobs.

> My circle of friends is small. I have made more male friends than female friends. I feel I have very little in common with Bajan females.

On the other hand, friendships with males were seen as much easier both to make and to regulate. The view that being different drew attention from Bajan men was cited by several of our female informants:

> Women here are not friendly. But the men *are friendly*!

> Making male friends is *not a problem*. But you have to be very clear what the score is. But this is accepted.

Race and Race Typing

In an earlier section it was observed that racial affirmation within a majority black society was seen as one of the positives in migrating and living in Barbados. But after living on the island for a little time, matters of race relations and racialized identities were reported as giving rise to considerable thought, and in some cases real surprise, among these returnees. Saliently, the narratives of the informants often referred to the incidence of what they described as 'black-on-black' or 'colour-class' racism, where skin tone affects the standing and treatment of people on a day-to-day basis. On other occasions when talking about race, the second-generation migrants spoke freely concerning what they frequently referred to as 'white-on-black' racism, stressing the essential privilege accorded to whiteness. Both of these aspects are covered in turn in the body of this chapter.

Commentators on the contemporary social scene in Barbados have stressed the persistence of elitist cleavages between the black and white populations (see, for example, Lowenthal 1972; Layne 1979; Karch 1985; Potter and Dann 1987; Potter et al. 2004). Western (1992), in his study of Bajans in London, talks of the remnants of a colonial racial hierarchy and attendant racism in Barbados. In the wider Latin American context, Martinez (1998; 2004) has bemoaned what she refers to as the bi-polar, 'black-white model' of racism, arguing that it discourages the perception of common interests among 'people of color', and thereby contributes to the maintenance of white supremacy.

A number of the young Bajan-British returnees expressed the view that in their eyes, Barbados was still very much associated with slavery, as is shown by the three extracts that follow. Thus, as noted earlier, the view was expressed several times that 'Barbados did slavery the best', in the sense of being most enduringly affected by slavery. This accords well with the view that slave conditions were particularly harsh in Barbados. On several occasions, this was projected onto present-day social structure by the second-generation Bajan-Brits:

> They said this was a calming island. They used to send them here to break them, to wear them down. It ain't change, man.

> Ever since slavery used to move around ... I think Bajans were the best at slavery as well. I think they have still got a slave mentality.

> No, black people here were the best at slavery. People hear about Haiti and Jamaica, but here (Barbados) they are just fully washed over.

One of the main reactions among the young Bajan-Brit migrants was that the level of racism is more of an issue in the Caribbean than they had anticipated before departing from the UK/North America. In particular, several were surprised at what they specifically described as the degree of 'white-on-black racism' (Martinez 1998 2004) in Barbados. This involved explicit reference to the hegemony of white commercial control in Barbados, and the acceptance of segregation between racial groups. In one case this even included mention of the influence of the plantocracy:

> I find people much more sensitive to race here than I expected. I mean in terms ... in Barbados, white-Bajans, black-Bajans, there is much more of an issue than I realized when I came here before. You see all of the businesses are owned by white Bajans. Most people in high places are white-Bajans. You know, that is the result of a plantocracy.

Another informant went so far as to make a direct comparison between the operation of racism in the UK and in Barbados, maintaining that:

> In a sense, I tease my father and say I have experienced more racism in Barbados than I ever did in England. But in a sense it is true.

while the operation of both 'white-on-black' and 'black-on-black' racism was specifically commented upon by another Bajan-Brit migrant:

> I find that there are two types of racism here. There are black racists and white racists. The 'haves' and the 'have-nots'. The blacks are the majority, but the whites still hold all the power.

Other commentators have noted the importance of colour-class differentiations in the Caribbean (see Potter et al. 2004). In fact, such distinctions are common throughout the Americas. Thus, Lancaster (1991) has given an account of how skin colour, race and racism operate in the context of Nicaragua. While the official political line is that racism does not exist, Lancaster demonstrates how dark skin is directly associated in public consciousness with poverty, backwardness and evil. In contrast, affluence, power, status and wealth are seen as being irrevocably bound up with whiteness.

The narratives of our informants showed all too clearly that our sample of recently arrived second-generation West Indians born in the UK were acutely aware of the continued existence and operation of the colour-class system within contemporary Barbadian society – that is, the social significance attributed locally to gradations in skin colour. As we have noted previously, this was frequently referred to as 'black-on-black' racism in the Barbadian context, in addition to the black-white mode just discussed (Martinez 1998, 2004; Lancaster 1991).

There was frequent direct acknowledgment among the Bajan-Brit second-generation migrants that skin colour remains directly associated with perceived social standing in Barbados:

> … what people say is true … that the lighter your skin, the more opportunities you would get.

> … but I can see they favour light skinned people.

Many narratives reflected on the 'local hierarchy of color' in Barbadian society (Western 1992: 48):

> Bajans here … if you are dark, you from the lower class and if you have a lighter complexion, you would get through here in life. 'If you are black you stay back', you know? Like in the duty-free stores: if you go in there as a black person, the cashiers don't notice you. They think that you don't have money to spend. And all of their skin colour is light too.

One interviewee linked this with naming, along with clear connotations of class and affluence:

> Barbadians go by colour here. Sometimes they even use nasty names that people would find offensive. And also there is a class-consciousness here.

> People who have made it look down on the ones who haven't. The way they talk to their employees in shops.

Yet another informant talked of the light-skinned sales assistants at Cave Shepherd, a leading Bridgetown department store:

> You know if you go into Cave Shepherd as a black person, you see only high brown people at the counters. And they get on like you can't pay for a thing.

One Bajan-Brit referred directly to the occurrence of what she specifically referred to as 'black-on-black' racism:

> I have observed racism, especially black on black. You know, they have this attitude: 'how dare you' or 'who do you think you are?'

A similar sentiment was expressed by another informant:

> I have seen the preconceptions that people have with dark skin, even in my own family.

As we have detailed elsewhere (Potter and Phillips 2006a), the possession of an British accent appeared to bestow an automatic symbolic lightening of skin colour on the relatively young Bajan-Brit migrants:

> ... still, half of Barbados think I am white, because I have an English accent. Most people when they meet me are surprised – and it is not even that I am black – but I am dark. They expect me to be fair.

It is notable that this informant talks in terms of distinctions such as 'white', 'black', 'fair' and 'dark'. This particular narrative serves to stress the instability of racialized identities among these second-generation migrants. Thus, this individual is taken to be white on the basis of accent, is self-professed to be 'dark' rather than 'black', but in the end, is expected to be 'fair'.

Several Bajan-Brits expressed their frustration with the Bajan obsession with gradations in skin colour, in a manner that directly paralleled Martinez's argument that it drives a wedge-between 'people of color'. In several instances, the general point was made that in the UK, if you are not regarded as white, then you are regarded as black. In general discussions with the Bajan-Brits there was an argument that the operation of the colour-class system in the Caribbean served to diminish what was seen as the unity of black people (Potter 2003a):

> I have lived next door to white people. But here the divide between black and white is strange. No white people down here, no white people on the bus. Even my boss, who is married to a black Bajan woman, tends to go out with more white people. You never see him anywhere with a bunch of black people. There is racism amongst blacks too. I know that from years ago. I used to work with a 'high brown' lady who would always call herself 'coloured' or 'light brown'. And I would always say, 'in the eyes of the white man you are black'!

Again, this narrative connotes the instability of racialized identities, varying from being 'high brown', 'coloured', 'light brown' to 'black'.

Indeed, according to the narratives of the informants, the operation of the colour-class system seems as strong as ever. Thus, one individual started by strongly refuting the tenability of black-on-black racism:

> There can't be black on black racism, 'cause we are all the same race ... I found that people don't seem to recognize that any shade of skin colour that is not white is black, which is what we grew up with, you know what I mean? You could be the clearest guy in England and you would get beat up just like me. I personally haven't been affected by it. I don't think that anybody has ever come to me with anything like 'you are too dark'.

But in presenting this negation, the informant ended up providing an anecdote that cogently exemplifies its operation:

> I have heard of some horrible incidents. Like a dark girl was working at a hotel. A very dark girl – nice personality – a waitress and they fired her basically because she was too dark. She was temporary or part-time staff. She used to work three or four days a week and she used to be very, very friendly – good personality. When I saw her one day she was almost crying. So I said: 'what's the matter?' She said that they sent her home. They wouldn't tell her, but one of the staff members said that they said she was too dark. And some of the tourists – well, not so much the tourists, but local Bajan whites, had objected to being served by her. She was a big girl, very dark. She had a couple of hairs on her chin. People here have this perception that if you are clear, you can get through in life, you can get a better job ...

Another Bajan-Brit migrant records the essential confusion that appears to surround her racialized identity within Barbadian society:

> No one would ever have considered me not black in England. But yet, when I came here ... I mean, absolutely amazing when people ... you know ... confusion!

Interestingly, where our Bajan-Brit migrants argued the case for changing patterns of social stratification, they almost invariably came to an ambivalent position, arguing that both class and race distinctions still exist. Thus, one informant commented that racism was:

> ... not as much in my face as I thought it would be. I think it is socio-economically. The blacks have their own version. It is very classist. You are not supposed to mix. My Mum thinks so too. I think it is something to do with the old generation. I have no Bajan white friends, only expats. I am offended more by people who act ignorant.

Another informant made the case for the existence of an increasingly class-based, as opposed to an entirely race- and colour-based, social ordering. However, in stressing the salience of class, the dimension of skin colour quickly finds its way into the argument:

> Yeah, not racist, *classist*, I would say. Well ... racist and yes, classist, two types of distinctions there. You've got your racist, where obviously the majority of blacks aren't earning, that don't have businesses and that sort of thing. Where the majority of whites or Asians or whoever you want to call them, are owning all the big businesses, have all the money and are in control. That I find racist. It is very big in Barbados. Then you have your class system where you have black people that do have a bit of money, 'cause they have managerial jobs, or work in certain fields – banking, whatever else. And they have money and they don't necessarily mix with, like, the working class, if you like – I suppose similar to England. I don't know, but it is the same sort of thing in England, but it is a shame to see black people doing it to themselves, if you know what I mean, and still not getting anywhere at the end of the day, 'cause you still don't own a lot of the companies. And I think that Barbados, because of its educational system and how it is, should be doing a lot better with its black community.

Indeed, this slippage between skin colour and class was frequently exemplified in the narratives of other second-generation migrants:

> I find this a very class-based society – more so than I expected. Very much into colour and quite separatist.

> You think you leave England to get away from prejudice, but it is the same thing here. I think it is more of a class thing, more than a race thing. I think the darker your skin is, depends on your class here.

> I think it is not about colour, it is about class here. I went through a lot of racism in England, 'cause I was darker than the rest.

Some of the intricacies of the arguments concerning race and class were well expressed by one informant, who is left to contribute the last word in this section. While this individual clearly stresses the contemporary salience of class over skin colour, she does so in a context where race is still seen as counting for a great deal:

> I knew Barbados was racist already, but everywhere you go there is racism and as black people we must learn to rise above it and live. It is a pain in the backside, but it exist everywhere you go in the world. Colour grading exists in Barbados. Where I would get through is on the premise that I am English. If you come from middle class Barbados, you would get through irrespective of your shade. … Racism amongst black on black is far higher than white on black. The reason why I would say that is that the interaction with a black person and a white person on an everyday basis is very, very remote. It only happens in a working capacity.

Accusations of Madness

In the pilot study we were taken aback when a recurrent theme emerged, with the interviewees reporting that they are frequently labelled as being 'mad' (Potter 2001a and 2000b, 2003a, Potter and Phillips 2002), and this was just as characteristic of the in-depth interviews we carried out as part of the main study. For example, one informant made it clear that this was all too frequent an occurrence in day-to-day interrelations in Barbados:

> I know that English being mad is a very common Bajan conception.

While another suggested that a basically philosophical attitude had to be adopted with respect to such frequently encountered comments:

> I was told that English people are mad and that English people come back and brought England with them. There is no point getting upset.

Saliently, as in the extract above, many of our informants made it clear that the accusation of 'madness' is frequently made openly and within earshot of them. This suggests that the description is on occasion being used as a social marker, a point that is picked up in detail subsequently:

> I was called 'mad' many a time. Where I was working once, it was even the topic of discussion in the lunch-room that English people are mad. So I asked: 'Am I mad'? And no one could answer me …

In another extract, madness among those who have lived in England appeared to be directly conflated with the incidence of mental ill-health among the Afro-Caribbean population in general (Reid-Galloway 2002):

> You hear a lot about English people being mad. I mean, even my relatives. I saw a documentary about black people being mad, and if I didn't know better, I would think the same thing.

Indeed, another informant expressed the view that such labeling is so prevalent that resilience is an essential prerequisite of living in Barbados:

> Don't come over here expecting a holiday-type experience. It is not the same. The attitude of people is quite negative – and be prepared to be called 'mad'. You have to be tough.

In several instances, the outward manifestation of an English accent seemed to play a part, or perhaps more accurately, appeared to be the auditory trigger for the madness accusation to be openly articulated:

> People would ask me about my accent and I would tell them I am from England. ... Then they would say that English people are mad – and returnees resent that!

> When they hear the English accent, they tend to treat you differently. I mean, they have been saying that English people are mad for years.

But why is the madness appellation seemingly employed so readily in relation to these young transnational migrants? In a detailed exposition (Potter and Phillips 2008) we interpret these accusations of madness as basically boiling down to five major sets of factors, namely: (i) explanations that relate to the historical-clinical circumstances concerning the incidence of mental ill-health among early West Indian migrants to the UK; (ii) madness as perceived behavioural and cultural difference; (iii) madness as 'othering', 'outing' and 'fixity'; (iv) madness as a pathology of alienation that is attendant on living in Barbados; and finally, (v) madness as the 'differential indoctrination' of migrants in the UK relative to the United States. The principal conclusion, therefore, is that the term 'mad' is being used in order to 'other' and 'fix' the second-generation migrants, in a manner that stresses their difference and leads to the follow-on that it is not necessary to listen to what they say, or to respond to what they suggest.

Conclusions

This chapter has focused on second-generation 'Bajan-Brits', who have 'returned to their ancestral home'. At the outset it was noted that several of our informants stressed the advantages of racial affirmation that were attendant on their 'return' to the Caribbean region. It was also shown that the second-generation Bajan-Brit migrants occupy an essentially privileged space within Barbadian society with respect to the employment market, finding it relatively easy to obtain jobs and to progress their careers. However, in the wider social and cultural realms it has been shown that their situation is far more ambivalent and contested. But both these privileged – and less privileged – spaces appear to be fundamentally predicated on the Britishness of the young returnees, which in a range of circumstances can be seen as giving rise to forms of 'token whiteness'. The Bajan-Brit returnees inhabit an economically privileged but uneasy space, involving them in countless contestations regarding their precise identities in the post-colonial context in which they find themselves in present-day Barbados (Potter and Phillips 2006). Thus, the research speaks of the liminal, hybrid and in-between racialized identities of these second-generation migrants. For example, most reported that they were acutely aware of the incidence of race typing and racism in their adopted home of Barbados, reacting not only to the existence of 'black-on-white' racism, but with a real measure of surprise to the occurrence of 'black-on-black' distinctions in society, involving the operation of colour-class gradations (see Potter and Phillips 2008). As a further instance, we report on the extreme problems that many of our informants experienced with respect to the formation of new friendship patterns in the Caribbean (Phillips and Potter 2008). Friendship patterns appear to be particularly difficult to forge for female migrants who reported serious issues relating to both workplace and sexual competition. Thirdly, in what appeared to be very much grounded within the local Barbadian socio-cultural milieux, many of the young migrants reported that they were frequently labelled as 'mad' (Potter and Phillips 2006). This 'madness trope' can be interpreted as a means of 'othering' and 'fixing' the returnees in a manner that means that they do not have to be listed to, or their advice heeded. It is in these ways that the second-generation migrants to Barbados stated that they experienced the kinds of feelings of alienation, anomie and hybridity that are implied in the evocative expression of being metaphorically suspended 'in mid-air' between two societies.

References

Conway, D. and Potter, R.B. (2007), 'Caribbean transnational return migrants as agents of change', *Blackwell Geography Compass*, 1(1): 25–45.

Gmelch, G. (1980), 'Return migration', *Annual Review of Anthropology*, 9: 135–59.

Karch, C. (1985), 'Class formation and class and race relations in the West Indies', in Johnson, D.C. (ed.), *The Middle Class in Dependent Countries*, Beverley Hills, CA: Sage, 107–36.

Lancaster, R. (1991), 'Skin colour, race and racism in Nicaragua', *Ethnology*, 30: 339–53.

Layne, A. (1979), 'Race, class and development in Barbados', *Caribbean Quarterly*, 25, 120–35.

Lowenthal, D. (1972), *West Indian Societies*, Oxford: Oxford University Press.

Martinez, E. (1998), *De Colores Means All of Us: Latina Views for a Multi-Colored Century*, Cambridge, MA: South End Press.

Martinez, E. (2004), 'Seeing more than black and white', in Anderson, M. and Collins, P. (eds), *Race, Class and Gender*, Belmont, CA: Thomson, 111–17.

Phillips, J. and Potter, R.B. (2003), 'The social dynamics of "foreign-born" and "young" returning nationals to the Caribbean: a review of the literature,' *Reading Geographical Paper*: 167.

Phillips, J. and Potter, R.B. (2005), 'Incorporating race and gender into Caribbean return migration: the example of second-generation "Bajan-Brits"', in Potter, R., Conway, D. and Phillips, J. (eds), *The Experience of Return: Caribbean Perspectives*, Aldershot and Burlington, VT: Ashgate, 69–88.

Phillips, J.R. and Potter, R.B. (2006), '"Black skins and white masks": post-colonial reflections on race, gender and second-generation return migration to the Caribbean', *Singapore Journal of Tropical Geography*, 27: 309–25.

Phillips, J.R. and Potter, R.B. (2009), 'Questions of friendship and degrees of transnationality among second-generation return migrants to Barbados', *Journal of Ethnic and Migration Studies*, 35: 669–88.

Potter, R.B. (2001a), 'Narratives of socio-cultural adjustment among young return migrants to St Lucia and Barbados', *Caribbean Geography*, 12: 70–89.

Potter, R.B. (2001b), '"Tales of two societies": young return migrants to St Lucia and Barbados', *Caribbean Geography*, 12: 24–43.

Potter, R.B. (2003a), '"Foreign-born" and "young" returning nationals to Barbados: results of a pilot study,' *Reading Geographical Paper*: 166.

Potter, R.B. (2003b), '"Foreign-born" and "young" returning nationals to St Lucia: results of a pilot study,' *Reading Geographical Paper*: 168.

Potter, R.B. (2005a), '"Young, gifted and back": second-generation transnational return migrants to the Caribbean', *Progress in Development Studies*, 5: 213–36.

Potter, R.B. (2005b), '"Citizens of descent": foreign-born and young returning nationals to St Lucia', *Journal of Eastern Caribbean Studies*, 30: 1–30.

Potter, R.B., Barker, D., Conway, D. and Klak, T. (2004), *The Contemporary Caribbean*, London and New York: Pearson/Prentice Hall.

Potter, R.B., Conway, D. and Phillips, J. (2005), *The Experience of Return: Caribbean Perspectives*, Aldershot and Burlington, VT: Ashgate.

Potter, R.B. and Dann, G. (1987), *Barbados: World Bibliographic Series*, Vol. 76, Oxford, Santa Barbara, CA and Denver, CO: Clio Press.

Potter, R.B. and Phillips, J. (2002), 'The social dynamics of "young" and "foreign-born" returning nationals to the Caribbean,' *Centre for Developing Areas Research Paper* (CEDAR, Royal Holloway, University of London), 37.

Potter, R.B. and Phillips, J. (2006a), 'Both black and symbolically white: the Bajan-Brit return migrant as post-colonial hybrid', *Ethnic and Racial Studies*, 29(5): 901–27.

Potter, R.B. and Phillips, J. (2006b), '"Mad dogs and transnational migrants?" Bajan-Brit second-generation migrants and accusations of madness', *Annals of the Association of American Geographers*, 96: 586–600.

Potter, R.B. and Phillips, J. (2008), '"The past is still right here in the present": second-generation Bajan-Brit transnational migrants' views on issues of race and colour class', *Environment and Planning Series D, Society and Space*, 26: 123–45.

Rhoades, R, (1979), 'From caves to main street: return migration and the transformation of a Spanish village', *Papers in Anthropology*, 20, 57–74.

Reid-Galloway, C, (2002), *'The Mental Health of the African Caribbean Community in Britain*, London: Mind Information Unit.

Springer, H,W, (1967), 'Barbados as a sovereign state', *Journal of the Royal Society of Arts*, 115: 627–41.

Western, J, (1992), *Passage to England: Barbadian Londoners Speak of Home*, London: UCL Press.

Chapter 6
Emulating the Homeland – Engendering the Nation: Agency, Belonging, Identity and Gender in Second-Generation Greek-American Return-Migrant Life Stories

Anastasia Christou

This chapter looks at the agencies, transformations and obstacles that intersect in the diasporic lives of second-generation Greek-American 'return migrants', which are characterized by both mobility and stasis. The returnees, who are mostly young, middle-class, highly-skilled and highly-educated professionals, imaginatively construct and implement their 'return project' as a *praxis* of their signification in terms of their identification, sense of belonging, as well as contribution to their ancestral homeland. In practice, however, they mostly encounter spaces of antagonism and exclusion (Christou 2006c). Such encounters become spaces of transformation in their changing (ethnic and cultural) identification and sense of belonging, which are filtered through their experiences of migrancy and diaspora. Furthermore, their life stories and narratives develop into a multidimensional context of agency, whereupon the verbalization and narrations of their transcultural-selves become a discourse to explore questions of identity and home that are remoulded along gender, ethnicity, class and generational lines (Christou 2006b).

These return migrant narratives can be characterized as situated subjectivities and acts of cultural performance of the migrant-self (Christou 2006a). The return journey is itself an act of nostalgic belongingness in search of an authentic ancestral homeland and relocation to a cultural space of 'Greekness' which embodies Greek ideals of traditional ethno-religious values (Christou 2003c). Through life history oral and written narratives, second-generation Greek-American 'return migrants' express their experiential, constructive, negotiated and imaginative notions of home and belonging in their diasporic lives. Furthermore, their state of migrancy and displacement becomes exacerbated throughout their settlement and adjustment processes as they confront new and 'foreign' circumstances in a modern Greece that is quite different from the one they imagined or had previously experienced in the distant past (Christou 2006c; Christou and King 2006).

The return migratory project is a multifaceted phenomenon that entails complex socio-cultural modifications of migrants' transnational experiences (Panagakos 2003; Christou 2006d). These processes and outcomes are reflective of agency but also indicative of how identifications are altered by mobility and settlement. Hence, while focusing on the narratives of return through encounters of 'home' and 'exile' in the ancestral homeland, the multidimensionality of generational, gendered, ethnic- and class-based processes of diasporic lives and identifications is unveiled. In this way the inseparability of agency and identity becomes profoundly apparent as the narratives illuminate the diversity of ways in which returnees' praxes both forge and reflect emerging facets of their identities.

Patriarchal representations of gender have a direct impact on women's lives and hence find their way into women's subjectivity and agency in how they describe themselves and construct the imagery of their futures (Tietjens Meyers 2002). Life stories, when recounted, can become narratives of resistance, particularly when they relate to how women can acquire self-determination for personal and cultural change. Life stories are an avenue of 'reading' the self and in this process a 'translation' of selfhood emerges. In this way women articulate their needs while enacting their own storied lives. By such means, women can make sense of change in their lives, and they can discover their capacity to voice and rewrite the micro-cultural contexts of their everyday lives.

However, it is important to note that the focus in this chapter is on the subjectivities of self-transformation through return migration and the meanings that we can derive from the act of relocation to the ancestral homeland. The framework that underpins such an analysis is that second-generation mobility is an act of transnational 'counter-diasporic belongingness' (Christou 2006d; King and Christou 2008). Moreover, the sense of belonging is very much influenced by the degree to which ethnic heritage and ethno-religious identification shapes the way the second generation negotiates their dual/hyphenated identity (Christou 2001). And, such negotiations reflect a non-linear perspective of a process of 'culturalization'; that is, development of the group's cultural awareness in the multi-ethnic spatial realm of the 'host' country. This is the kind of conceptual framework that Kourvetaris (1999: 149) advocates when he emphasizes that, 'while we are talking about ethnic diversity and experience, we must also keep in mind that the Greek American experience is itself diverse. While most Greek Americans share similar experiences as members of ethno-religious communities, we cannot speak of a homogeneous Greek American community. Indeed we can speak of a generational Greek American subculture by looking at the first, second, third, and subsequent generations and the class dynamics of these generations'.

Dynamics of gender are also important in understanding expressions of identification, as Orfanos (1999: 281) claims:

> levels of Greekness surely exist on a continuum, as do gender arrangements. Researchers in the field of Greek American Studies have a very hard time defining who is Greek American. Right? Wrong! Greeks in Greece or Greek-Americans in America also have considerable difficulty in defining themselves, their ethnicity, and their culture. And so is the case for gender.

Male and female norms rooted in the ethnic context are very much relevant to how 'roles' are defined and stereotypes reproduced. For instance, patriarchal prejudice and pride are directly responsible for the low status of women because they are institutionalized (Topping 1983).

The discussion that follows will draw on the written, life story narratives of two female participants, who contributed reflective journals[1] which recorded their experiences of relocation and settlement in their ancestral homeland. I situate this discussion in the context of 'return migration' as a 'site of struggle' in negotiating belongingness and identification and grounding the narratives within particular conceptual 'triangulations' of gender. By so doing, I intend to analyze how individual narratives intersect with lived experiences in the 'homeland', while treating the context as an interconnected cultural space. Narratives that 'translate' past cultural experiences while explaining the present are understood as both memory-making and meaning-making activities of remembrance and activity. Accordingly, they serve to illuminate levels of consciousness on the transactional and structural impacts of mobility and underscore the agency of 'return migrants' as active actors. Theoretically, questions on migrancy as both an individual diasporic condition and a collective sense of identity and the role of gendered identifications in transitional contexts are considered. Moreover, narratives are a process of 'self-interpretation' as individuals speak for, about and through, their life stories by the use of personal testimony. In order to make sense of the world surrounding them, participants deploy a series of cultural constructions in

---

1 'Lydia' and 'Alexandra' are pseudonyms used for the two female participants. Although the women had participated in an earlier project (2001–2003) in providing oral and written accounts of their life stories, the narratives used in this chapter were maintained between the summer of 2006 until the spring of 2007 in addition to personal informal conversations that took place on several occasions during this time-frame.

Methodologically this opportunity served to re-address issues of their relocation decision and their narratives reaffirmed earlier contributions and reconfirmed initial findings. Both women are in their mid-late 30s and made a conscious decision to relocate to their ancestral homeland about six years ago while their immediate family (parents and siblings) have no intention of moving to Greece.

such 'interpretations'. In this way, these narrated 'personal testimonies' mediate subjective understandings of their social world(s).

Migrancy, Diaspora and Return: Gendered Paths and Transitional Spaces

The burgeoning literature on transnationalism (Basch et al. 1994; Glick-Schiller et al. 1992; Levitt et al. 2003; Smith 2001; Smith and Guarnizo 1998; Vertovec 1999, among others) has recently been confronted by a reconsideration of the notion of migrancy. Such reconsiderations of migrancy (Christou 2006a; Harney and Baldassar 2007) aim at interpreting social and cultural practices characterized by transnational relations and they identify three essential elements:

1. the interconnections between movement in space and time in transnational practices;
2. the need to de-emphasis the nation and give consideration to the agency of migrants;
3. the consideration of power inequalities, not only through the nation-state but also through discourses of difference and otherness.

Undoubtedly, migrancy is a central metaphor of the contemporary world; both as an emerging physical phenomenon of mobility with its implications for everyday life, but also because the intellectual pursuit of studies of migrancy is revealing the complexity of our social worlds (Chambers 1994). Furthermore, by unveiling the gendered paths of relocation and settlement, we focus on the social and cultural constructions of gender and hence the relationship between constructions of maleness and femaleness (Strathern 1988) when referring to a culture's assumptions about the differences between men and women, the roles they play in society and what they represent (Domosh and Seager 2001).

Journeys of 'return' migration are not just acts of physical relocation, simply involving territorial shifts. According to Wilding (2007: 332), it is important to note that:

> in such journeys, it is not just the physical movement, or corporeal 'travel', that is important. What I argue here is that, more important than the traversing of geography in these stories is, rather, the movement of the mind or imagination, enabling a shifting of cultural and perceptual frameworks.

One of the most interesting paradoxes of transnationalism is what Hannerz (1996) identified as the denial of significance of national boundaries on the one hand, but the retention of the nation as a central category of analytical significance in 'cross-border' relationships between one another. In such transnational relationships the links are multiple; practical and practiced, experienced and experiential, but also virtual and imagined.

I use the narrative journals in two overlapping ways: both as a reflective device through which intimate inner thoughts and feelings are expressed, as well as a textual device through which those reflective thoughts and feelings are mediated in a neutral space. One of my objectives in doing so is to set up this neutral space so participants could explore the re/negotiation of 'feeling' and 'being at home', while recognizing the zones of safety, comfort and belonging that could possibly be 'identified' as 'home'.

Through storytelling I examine glimpses of the women's 'diasporic ancestral homeland' lives – their stories provide a view of life lived in the microcosm of their everyday life in the ancestral homeland as second-generation Greek-Americans. Above all, these glimpses of a lived life are fragments of diasporic subjectivities, narratives of self-searching and gendering stories of 'fluid identification'.

Obviously, contradictions could emerge in the ways that women discuss their lives and identities, but this is precisely the nexus of their reflective subjectivity. Elsewhere (Christou 2003b, 2003c) I have explained how second-generation Greek-American women perceive their identities as wife, mother, housewife and homemaker and within these identities, their professional selves also unfold and hence their aspirations and realities move beyond those that are culturally prescribed. Female (return) migrant identities are subjectivities shaped by ethno-cultural signifiers which are also embodied in a lived reality, (re)negotiated in a dynamic process. Women's narratives express the ways of their resistance: they are performed and situated in their everyday lives. Between conformity and resistance the narratives are discourses that create varied meanings of 'home' and 'belonging'. Although the process of relocation represents uncertainty and instability, it also signifies a 'beginning' with everyday participant practices becoming, then being, purposeful 'agency'.

Stories of Self – Self-Narrated Stories of 'Home'

The re-theorization of 'diaspora' as an analytic term emerged quite energetically in the last decades. As Kandiyoti (2003: 10) notes, 'diaspora theories of the 1980s and 90s attempted to rescue the heuristic concept from its former, predominantly Jewish significations as well as its association with disaster and separation'. Most influential have been the works of cultural theorists such as Paul Gilroy, Stuart Hall, and James Clifford, for whom 'diaspora' came to denote, among other things, a productive multiplicity, and not only suffering and disconnection. Contrasting diasporas to immigrant communities, in his influential article, James Clifford (1994: 311) has maintained that 'Diasporist discourses reflect the sense of being part of an ongoing transnational network that includes the homeland, not as something simply left behind, but as a place of attachment in a contrapuntal modernity'. In one of the leading works conceptualizing diaspora, Paul Gilroy (2003: 10) uses W.E.B. Du Bois' notion of '*double* consciousness' applied to the diasporic situation, stating: 'There is a general sense that the double or multiple attachments afforded by diasporic modes and sensibilities lead not to disunity, diffusion, and treason, as nationalist discourses would maintain, but to a gainful multiplicity of perspectives, languages, and knowledges.'

Diaspora has been viewed as a process in deconstructing boundaries (Mavroudi 2007), but also as an important component in the discourse of boundary maintenance versus boundary erosion (Brubaker 2005). Furthermore, the diasporic condition has been problematized beyond the first generation in a continued state of migrancy as a 'counter-diasporic' act that reverses the scattering, namely through the return migration of the second generation (King and Christou 2008).

By keeping a journal of their most intimate thoughts and reflections in relation to their relocation, the women found that this was a path towards self-interrogation which was at times helpful in liberating some of their inner frustrations and coping with the obstacles or the stimuli to such anger. But during other occasions, the practice of facing their complex everyday life in a reflective mode was confusing. Lydia explains her feeling of 'being unsure of just where or how to fit it into the theme' of diaspora and (return) migrant life. In her journal she develops this further:

> My understanding is that I should write about my experiences of being a (second-generation) Greek-American woman living in Greece. Initially, I was sure that I could effortlessly write volumes on this specific topic, especially since I know there are no 'rules' and that it can just be free-association and informal. Sounds like just my thing. Then of all things, I got hung up on the adjectives, believe it or not. What I mean is it's difficult for me to separate them. Is what I experience due to the fact that I grew up in the US, is it due to the fact that I have Greek roots or is it due to the fact that I am a woman? Most probably it's all three combined but I have not really focused on them in combination before.
>
> The other problem I have in relation to this distinction is that it places me in a category – a category I am supposed to fit into and identify with and therefore relate to others who are in that same category and herein lies the problem: I have never been able to do that and I still can't. Over the years, I have come to realize that this has nothing to do with what continent I live on or with the others who are in the category that is supposed to represent me. It simply has to do with me and my aversion to associations or groups of people though I have given it my best effort.

Lydia poses some pertinent questions: on the one hand, do subjectivities of the self emerge based on particular experiences, that is, do we make sense of the self as shaped by one's 'lived' and 'living' life, or do social, cultural and ethnic categorizations impact the way we actually give meaning to such experiences? Furthermore, how much 'control' do individuals have over such categorizations? Do they have to 'fit into' distinct categories and hence situate their identities within such representations? And, finally, do all identifications have to adhere to a collective sense of association? For one thing, the very act of relocation positions particular individuals in specific mobility paths of a 'return migratory' nature, thereby rendering them as a collectivity of 'return migrants'.

By exploring the expressive representations of self in life-story narratives (Chamberlain 1997; 2000), we can make sense of participants as active agents who reflect on their life trajectories. One, Alexandra, offered these ambivalent reflections:

> I have been in Greece for five years now. It feels, both, like a long time and a short one. I came here thinking that I would try it out for a year and then return to NY, but it turns out that one year may turn out to be a lifetime. I am still torn about what I should do. On the one hand, I have gotten used to living here and feel settled in, but on the other I feel like I belong back in NY with my family. The truth is I like my life here, but it is full of guilt for leaving loved ones behind.
>
> I have been coming to Greece every summer of my life. I have traveled back and forth so many times, yet on the day when I landed on August 18, 2001, I had a feeling of terror in me. As the plane approached the airport tears began to well up in my eyes. I felt sick. This was definitely not a vacation. The first thing that I thought of was 'what are you doing?' To be honest, that feeling has never left me.
>
> When people ask me why I came here I usually answer that I wanted to see how life was, but I am not sure if that's even true. In doing some self-exploration lately I have tried to figure out if there is any other reason why I am here, but I don't think I have come up with anything. To be honest I really don't know why I came. Maybe it was a subconscious decision, and the more I think about it I am sure it was, but I would really like it to come to the surface because I am curious myself as to why I am here. The only thing is that I always felt 'something' pulling me here. However, that 'something' is still unidentified. Maybe it's the feeling I have here. Not the everyday feelings, but the general feeling. In general I feel like I am more in my element here. I can't explain exactly what I mean, it's the only way I can describe it. Maybe it's because I feel more free and independent. I think that in NY I feel more confined and suppressed. Here I don't have that feeling, except for professionally. Professionally I feel more limited here. I have fewer options. In NY I felt that if I wanted I could do anything.

Alexandra's narrative is indicative of the series of binaries and dichotomies that the relocation involves, namely the 'here and there' that do not seem to compliment or complete each other. Alexandra talks about feeling more comfortable in the ancestral homeland while at the same time she describes in great detail all the uncomfortable circumstances there which she has been forced to endure. Yet, in her country of birth and long term residence, the US, despite the abundance of opportunities, she feels restricted and repressed. Furthermore, to add to the 'cultural schizophrenia', participants are confronted not only by an 'ethnicized' or gendered categorization of their Greekness, but also by a classed-based classification:

> Here in Greece I live in Kallithea. It is a large *demos* with a multi-ethnic population. It is also considered a lower-class area. At least, that is the feeling I have when I tell people where I live. More than once, when I've told people that I live in Kallithea they give me a very odd look. As if 'you live where'? Eek. Usually these are people who live in the northern or sea-side suburbs which are also more affluent. I think that I have gotten used to it now, but in the beginning it was strange for me. I felt judged and looked down upon. I guess it still bothers me a bit and I am self-conscious because even now I don't usually divulge the info so readily. It seems to me that class separation is much more prominent here in Greece than it is in the US. I know I haven't been exposed to the whole population of Greece, but I feel that people usually associate with their own socio-economic class. I don't really appreciate this, because living in a city like NY I didn't really feel the separation that much. Of course, class difference exist there too and I am sure Trump doesn't hang around with the local hotdog cart guy, but I think it's much more likely than an average doctor here in Athens associating with the kiosk owner.

There seems to be a sharp difference in how social geographies of everyday life are perceived and experienced in relation to Greekness and Americanness. Although class-based differences are very much visible and apparent in the US, this does not seem to occupy the individual imaginary in ordinary life, whereas in Greece, conspicuous consumption and residential patterns become markers of class identification. This creates a psychological divide, especially when participants feel a sense of being looked down upon when their lifestyle and material possessions do not conform with the image of the successful and wealthy Greek-American according to the natives' categorization. Suddenly, class matters for the returnees.

Negotiating the Nation – Locating 'Belonging'

In locating a sense of belonging through a sense of place, dislocation through migration and diaspora often exacerbates the agonizing longing for the 'home(land)', since everyday life becomes saturated by ethnic signifiers in order to 'preserve' the nation in the absence of the nation-state. This process clearly 'maps the vicissitudes in the relationship between ethnicity and diasporicity' as 'the case of Greek America demonstrates that the formation of an ethnic identity has been entangled with narratives about national belonging and transnational origins' (Anagnostu 2003: 280–81). This is exemplified by the inevitable multi-sided experiences of diaspora, the spatial imaginaries and the temporal events that emerge in the intersections of 'home', memory and everyday life. That is, homing experiences of migrants involve both the materiality of everyday life that is both imagined and lived. The desire to (re)connect memories with objects is a signifier of the process of incorporating the 'nation' into everyday life and hence the migrant's subjective identification with a cultural self that embraces the 'ethnos' of Greekness. Lydia's lengthy narrative below presents the stages of such an enactment of the cultural self:

> If There Were No Music …
> I realized recently that I have stopped listening to music entirely. Well, I admit that sounds a bit drastic. What I mean is that I recently stopped listening to music on the bus ride to work, while at work when I am doing admin tasks, while I do household tasks and whenever I have down time at home. Having thought about this new phenomenon, I had to stop and ask myself why this has happened. Could it be that I have just exhausted all of the CDs in my music collection, even my most beloved homemade compilations? Not sure if that's precisely it. So then, what?
>
> When I first moved to Greece nearly four and a half years ago now, I had only two suitcases to my name and no apartment yet. The first month I was here, when I was staying with my aunt, my cousin (her son) gave me a boom box as a gift. I had brought only a few of my most beloved CDs with me initially, but of course had nothing with which to play them, except for my Discman but I am of the belief that music is meant to fill a room to really be enjoyed. He said, 'Music is vital; it provides companionship' and he said it like he knew something from experience and in advance, which in fact he did. I told him I completely agreed and was both touched and grateful.
>
> It was when I moved into my first place, however, when I realized that the music I had brought with me really was one of my few lifelines. In a place with no furniture but a cot, two cheap chairs, a small cheap table and with a view of a gargantuan grey cement building across the street, that's not very surprising. Not to mention that I of course did not have a phone yet. I

> knew all the songs by heart, most of them ballads that I enjoy to sing along to when the mood strikes. And they indeed provided companionship with lyrics that speak to my soul. My soul, yes, which was elsewhere, not with me at all. Therein came the flip side of things – the more they comforted me, the more they brought me pain. *The pain that comes from loss and longing of people and places, the pain that comes with memories of good times and bad all in a place that is now worlds away.*[2]
>
> I continued to listen because I needed that connection. And I came to realize that this music is part of who I am – *the real me, as opposed to this new other me – the stranger*. As time passed on, I began to adjust, to make connections to people and places. Familiarity set in. I fell in love again. I got married. I made new friendships while hanging on to the old. Slowly the music took on additional memories, linking my *old self* to my *new self* and blending more layers of experience. What I still missed regarding music was sharing it with my best friend, as she is one (perhaps the only) person in the world who has musical tastes almost identical to mine and whom I believe is in love with music itself. For this, there was no substitute. I also greatly missed hearing live music the way I used to in Minneapolis – everything from seeing my favorite famous bands and local bands to dancing to live blues in a bar. These are things one somehow has to *learn* to live without.
>
> Having thought all this through, I still am not sure I have come up with an answer to my initial question. Instead more questions arise: could it be that I have found substitutes to get me through those tasks and moments – like conversing with others or just listening to my thoughts instead? Am I too caught up in the daily grind of routine and the rhythm of life? Is it that I no longer need the lifeline in this medium? Or have I gotten so completely removed from all that constitutes the *real me* and all the small things I used to enjoy doing that I am simply adrift in a *vast abyss* in which not even music can get me through?

Lydia's long narrative excerpt encapsulates the struggles and successes involved in the project of relocating to the ancestral homeland. In between those stages from struggle to success (Christou 2001), there are multiple layers of what Lydia describes first as pain and trauma, then effort and compromise to reach a level of fulfilment and ease during the successive transformations of the self that follow. So relocation is a project of the self, it is an identity project (Christou 2006d), a deeply ontological and existential project (King and Christou 2008) and certainly a social project of homecomings that are, for the most part, unsettling paths of return (Markowitz and Stefansson 2004).

> I can't say that I felt sooooooo welcome by the girls at work. Actually that's not true. When I first started everyone was so nice and friendly and sweet

2 Italics used in narrative excerpts are mine for emphasis.

> and courteous. I can go on and on. However, in time, things changed quite a bit. I felt a bit like the outsider. It may have nothing to do with the fact that I was from the US, but I really got the feeling it did. That's another issue here. Local people do not see me as one of them. Anyone I met put me in a different category. 9/11 didn't help the situation either. Some people couldn't wait for the opportunity to rub it in my face. They acted as if the whole thing was my fault, or if they could get it in the conversation somehow they would. And of course the comments were not very favorable. I've heard things like 'the whole country is going to sink', 'good for them', 'I was so happy when the towers fell', and others. In stating all these feelings I have, I sometimes wonder again, what am I doing here, what is keeping me here????????'

It seems that perhaps the particular stage in the life course is very pertinent to why, for example, Alexandra and Lydia decide to 'break away' from the scripted narrative of Greek-American life in the United States. Yet, even in such 'radical' cases of autonomy in one's narrative script, the significance of belonging in a cultural context is apparent. As Anagnostou (2003: 295) emphasizes;

> in view of fragmentation and the scattering of meaning, fluidity, and contingency – all in a context of rampant cultural commodification – coherence, stability, and continuity have become the central ingredients of narratives seeking to anchor individuals in enduring structures of collective belonging.

Conclusion

Return stories are for the most part narrations based upon idealized portrayals of an imagined and imaginative ancestral homeland, which is locked in a temporal domain that no longer exists; as rapid modernization and historical transformation both attest to its newness as well as its territorial heritages (Christou and King 2006). Yet, despite the hardship and trauma that the relocation entails, both participants decide to remain in the ancestral homeland voluntarily, as acts of autonomous decision-making. Lydia decides to marry a Greek and settle in Greece while an only child and her parents still remain in the US. Alexandra keeps attempting to formulate a personal intimate relationship but has been unsuccessful so far. She too, remains in Greece while her parents and two other siblings live in the US. The last time we spoke she was actually in the midst of arrangements with architects and engineers in the process of building a 'home' in Athens: a 'home' of bricks and mortar, in which the built environment encapsulates the crux of her dreams and aspirations.

But 'home' is both a space and place, a time and a stage in one's life. Home is as much fluid as it is rigid, it is flexible and complex. It seeks to ground and localize, but it is also an integral part of a world of movement, it is relative and contested, a site of ambivalence and a source of anxiety. Home as a concept that raises issues of belongingness can become complicated and difficult to deconstruct and even to contextualize and situate. It may trigger memories, trauma, indifference and evoke struggles over selfhood and nationhood. The irony is that instead of giving people a sense of stability, balance and relief, the search for 'home' can be agonizing and a source of pain when displacement and dislocation occurs, as illustrated in the following excerpt:

> I guess that is what I feel about people here, that they are strangers. The overall feeling I have is that I don't belong. I don't have a sense of community like I did in NY. When I go to NY I feel like I am part of something. Here, I don't have that feeling. Lately though, I have tried to become involved with American organizations here in Greece. Being around those people (my people as I like to say) does give me a sense of community. We have something in common, we understand each other. Greek people don't make me feel like I am the same as them. They always point out differences, sometimes they are even positive. Some people state that those who were brought up outside of Greece have better ideals and values, that we are a better quality people. But usually my experience had been (ever since I was little) that I am the 'Amerikanaki'. The term itself has demeaning connotations due to the 'aki' ending. I, myself, cannot explain the meaning of an 'aki' ending, but it is put on a word to give a derogatory meaning.

Studies on the psychological well-being of Greek-American families, including their attitude toward therapy – but also sources of distress in response to immigration and acculturation – point to the very strained structure of such families in contemporary times. More specifically;

> Some problems encountered by Greek American families result not only from immigration, but also from the endemic, traditional patterns of organization and other norms and customs that regulate family life. First-generation Greek Americans uphold family traditions that are more conservative than those of second-generation Greek families. The first generation lives a marginal existence between two worlds: the Greek culture that they endorse but left behind no longer exists, and the American culture they live in is not one they will ever fully join. (Tsemberis 1999: 206)

The second generation seems to encounter another marginal space during the ancestral homeland return, namely, the one that exists between the pragmatic Greekness of their conservative-traditional life in the US and the imaginative Greekness of the (post)modern-progressive life in Greece. They are troubled primarily because they cannot bridge the rift between these two worlds and hence live a life in the margins. Furthermore, they themselves are marginalized by the native residents in their desperate search for a 'community' sense of belonging that will create a cohesive expression of support.

Lydia and Alexandra are cases that exemplify a sense of autonomy and independence in their adult decision-making processes. This is not always the case. In many instances, immigrant parents feel the need to maintain control by 'imposing what profession to choose, in which geographical area to live, and whom to marry. Daring daughters who decide their own course in life are rejected .... This may partially explain the numerous "cut-offs" in Greek families' (Hibbs 1999: 226). In a sense, one can only speculate if the ancestral homeland 'return' is an alternative form of 'escape' for some women: perhaps a multifaceted 'escape' from a rigid family structure and an inflexible social structure. Indeed, the relocation decision may be a combination of outcomes when socio-cultural 'externalities' and 'internalities'[3] interact. As a result, the exploration of an ancestral relocation could very well be a framework of conceived autonomy on the part of the second generation, but in reality one of perceived constraint. As for parental compliance with control, (especially when it comes to female members of the second generation):

> one can say that this inability to accept the daughter as an individual who has her own mind, goals, and search for individuation may have its roots in a multigenerational pathology through learning, which, in fact, is very difficult to eradicate. ...This rigidity, inability to change, and lack of openness can probably be attributed to their own difficulties with separation issues and fear of abandonment (rather) than to multigenerational repetition of cultural principles, since they themselves had the opportunity to be exposed and learn differently (Hibbs 1999: 226).

Lydia's and Alexandra's narratives epitomize the salience of notions of home, identity, place and belonging as constitutive elements of the cultural geography of second-generation ancestral homeland return (King and Christou 2008). Narrative journeys are journeys of self-discovery as much as the journey of relocation itself. The very act of relocation is a 'performative' action during which migrant agency is entangled within multiple stories of the 'home' and 'host' countries: the resultant narratives therefore range from the micro familial to the macro ethnic in context. But ultimately, these are all

3 In a study conducted by Constantinou and Harvey on 'Basic Dimensional Structure and Intergenerational Differences in Greek American Ethnicity', the authors found a two-dimensional structure underlying Greek-American ethnicity: one they called externalities (that which pulls the Greek American toward their place of origin), and the other they termed internalities (that which binds Greek-Americans together as a community) (Kourvetaris 1999: 155).

personal stories of 'subjectification' of the self in search of meaning, stability and grounding in time and space. Built into this journey is the multiplicity of ambiguity that mobility and the state of migrancy bring to transnational lives. Both Lydia and Alexandra articulate feelings and meanings of being, becoming and belonging, but also the struggle toward achieving closure in these complex processes in order to eventually resolve the trauma of diaspora. It is evident from their stories that both participants have quite intricate diasporic histories, as their parents and grandparents have multiple mobility experiences which have shaped their families' migration narratives and identities. Ultimately, their relocation to the ancestral homeland is in a way the therapeutic step in reconciling the rift that is beyond territorial and which is very much ontological. This chapter has, therefore provided a glimpse into such complex, yet often quite ordinary stages, in a diasporic existence; as reflected in two participants' return to their ancestral homeland – Greece.

References

Basch, L., N. Glick Schiller and C. Szanton-Blanc (1994), *Nations Unbound: Transnational Projects, Postcolonial Predicaments, and Deterritorialized Nation-States*, New York: Gordon and Breach.

Brubaker, R. (2005), 'The "diaspora" diaspora', *Ethnic and Racial Studies* 28(1): 1–19.

Chambers, I. (1994), *Migrancy, Culture, Identity*, London and New York: Routledge.

Chamberlain, M. (1997), *Narratives of Exile and Return*, Houndsmills: Macmillan.

Chamberlain, M. (2000), 'The global self: Narratives of Caribbean migrant women', in Cosslett T, Lurry, C and Summerfield, P (eds), *Feminism and Autobiography: Texts, Theories, Methods*, London and New York: Routledge, 154–66.

Christou, A. (2003a), 'Persisting identities: Locating the *self* and theorizing the *nation*', *Berkeley Journal of Sociology: A Critical Review*, Special issue: *Nationalisms: Negotiating Communities, Boundaries, and Identities*, 47, 115–34.

Christou, A. (2003b), 'Migrating gender: feminist geographies in women's biographies of return migration', *Michigan Feminist Studies*, special issue: *Gender and Globalism*, 17: 71–103.

Christou, A. (2003c), '(Re)collecting memories, (Re)constructing identities and (Re)creating national landscapes: spatial belongingness, cultural (dis)location and the search for home in narratives of diasporic journeys', *International Journal of the Humanities*, 1: 1–16.

Christou, A. (2004a), 'Human rights and migration – Theoretical reflections and empirical considerations on issues of social justice, ethics and

development: towards a redefinition of civil society', in Kyriakopoulos, V. (ed.), *Olympia IV: Human Rights in the 21st Century: Migrants and Refugees*, Athens: Sakkoulas Publishers, 217–25.
Christou, A. (2004b), 'Reconceptualizing networks through Greek-American return migration: constructing *identities*, negotiating the *ethnos* and mapping *diasporas* – theoretical challenges regarding empirical contributions', *Spaces of Identity*, 4(3): 53–70.
Christou, A. (2006a), *Narratives of Place, Culture and Identity: Second-Generation Greek-Americans Return 'Home'*, Amsterdam: Amsterdam University Press.
Christou, A. (2006b), 'Crossing boundaries – ethnicizing employment – gendering labor: gender, ethnicity and social capital in return migration', *Social and Cultural Geography*, 7(1): 87–102.
Christou, A. (2006c), 'American dreams and European nightmares: experiences and polemics of second-generation Greek-American returning migrants', *Journal of Ethnic and Migration Studies*, 32(5): 831–45.
Christou, A. (2006d), 'Deciphering diaspora – translating transnationalism: family dynamics, identity constructions and the legacy of 'home' in second-generation Greek-American return migration', *Ethnic and Racial Studies*, 29(6): 1040–56.
Christou, A. and King, R. (2006), 'Migrants encounter migrants in the city: the changing context of 'home' for second-generation Greek-American return migrants', *International Journal of Urban and Regional Research*, 30(4): 816–35.
Clifford, J. (1994), 'Diasporas', *Cultural Anthropology*, 9(3): 302–38.
Constantinou, S. and Harvey, M.E. (1985), 'Basic dimensional structure and intergenerational differences in Greek American ethnicity', *Sociology and Social Research*, 69(2): 234–54.
Domosh, M. and Seager, J. (2001), *Putting Women in Place: Feminist Geographers Make Sense of the World*, New York and London: The Guilford Press.
Glick Schiller, N., Basch, L. and Blanc-Szanton, C. (1992), *Towards a Transnational Perspective on Migration: Race, Class, Ethnicity and Nationalism Reconsidered*, New York: New York Academy of Science.
Hannerz, U. (1996), *Transnational Connections: Culture, People, Places*, London: Routledge.
Harney, N. and Baldassar, L. (2007), 'Introduction: migrancy and tracking transnationalism', *Journal of Ethnic and Migration Studies*, 33(2): 189–98.
Hibbs, E. (1999), 'Separation-individuation issues in the Greek American mother-daughter dyad', in Tsemberis, S.J., Psomiades, H.J. and Karpathakis, A. (eds), *Greek American Families: Traditions and Transformations*, New York: Pella Publishing, 223–35.

King, R. and Christou, A. (2008), 'Cultural geographies of counter-diasporic migration: the second generation returns "home"', working paper, Sussex Centre for Migration Research, University of Sussex, UK, in press.
Kandiyoti, D. (2003), 'Multiplicity and its discontents, feminist narratives of transnational belonging', *Genders* 37, http://www.Genders.org.
Kourvetaris, G.A. (2002), 'The futuristics of Greek America', in Orfanos, S. (ed.), *Reading Greek America: Studies in the Experience of Greeks in the United States*, New York: Pella Publishing, 145–66.
Levitt, P., DeWind, J. and Vertovec, S. (2003), 'International perspectives on transnational migration: an introduction', *International Migration Review*, 37(3): 565–75.
Markowitz, F. and Stefansson, A.H. (2004), *Homecomings: Unsettling Paths of Return*, Lanham, MD: Lexington Books.
Mavroudi, E. (2007), 'Diaspora as process: (de)constructing boundaries', *Geography Compass*, 1(3): 467–79.
Smith, M.P. (2001), *Transnational Urbanism: Locating Globalization*, Oxford: Blackwell.
Smith, M.P. and Guarnizo, L.E. (1998), *Transnationalism from Below*, New Brunswick, NJ: Transaction Publishers.
Strathern, M. (1988), *The Gender of the Gift: Problems with Women and Problems with Society in Melanesia*, Berkeley, CA: University of California Press.
Topping, E. (1983), 'Patriarchal prejudice and pride in Greek Christianity-some notes on origins', *Journal of Modern Greek Studies*, 1: 7–18.
Tietjens Meyers, D. (2002), *Gender in the Mirror: Cultural Imagery and Women's Agency*, New York: Oxford University Press.
Tsemberis, S.J. (1999), 'Greek American families: immigration, acculturation, and psychological well-being', in Tsemberis, S.J., Psomiades, H.J., and Karpathakis, A. (eds), *Greek American Families: Traditions and Transformations*, New York: Pella Publishing, 197–222.
Vertovec, S. (1999), 'Conceiving and researching transnationalism', *Ethnic and Racial Studies*, 22(2): 447–62.
Wilding, R. (2007), 'Transnational ethnographies and anthropological imaginings of migrancy', *Journal of Ethnic and Migration Studies*, 33(2): 331–48.

PART 2
Young and Youthful Return Migrant Experiences

Chapter 7
Back to Hong Kong: Return Migration or Transnational Sojourn?

David Ley and Audrey Kobayashi

This chapter reconsiders the meaning of return migration for Hong Kong-Canadians in a globalizing world of growing transnational practices. In this transnational era, their international movement is better described as continuous rather than completed, with migration being undertaken strategically at different stages of the life cycle. Strategic switching between an economic pole in Hong Kong and a quality of life pole in Canada identifies each of these transnational spaces as separate stations within an extended but unified social field. Return is often an episode of transnational sojourning, and more rarely the conclusion of a migration cycle. The narratives and experiences of returnees tell the stories of these temporary migrations and circulations.

A Representative Vignette on 'Return'

In August 2004 the political candidacy of Albert Cheng, host of a feisty open-line radio programme in Hong Kong, attracted prominent and simultaneous media attention in Canada and the Special Administrative Region. Mr Cheng is a leading media figure and his bold pro-democracy stance has become both a *cause célèbre* in Hong Kong and also a source of considerable tension with his broadcaster, who is nervous about recrimination both from organized crime and from the Beijing government. His decision to seek elected office in the Hong Kong legislature maintained a flamboyant public persona and also sustained high visibility for the pro-democracy movement in the September elections. But the candidacy also received attention on the front page of Canada's leading daily newspaper, because Mr Cheng, like tens, probably hundreds, of thousands of fellow citizens is a returnee to Hong Kong with a Canadian passport.[1] The newspaper's China correspondent was fully alert to the transnational content of the story as Mr Cheng told him that he was now

1 See York (2004). The same day's edition of the Hong Kong English-language daily, the *South China Morning Post*, contained no less than eight stories with references to Mr Cheng.

fighting in Hong Kong for liberal values he had learned in Canada: 'I have to stand up against violence and against any evil force that wants to shut me up ... This is a Canadian value. It's something I learned in Canada' (York 2004: A10). Moreover, he was obligated to renounce his Canadian citizenship as a requirement for running for office in Hong Kong, a step he had found to be 'a very serious and emotional decision'. But despite this heavy sacrifice and commitment to a long-term political project in China, he was not abandoning his transnational lifeline. 'I still consider myself a Canadian and one day I will retire in Canada and apply for my citizenship again' (York 2004: A1).

We argue in this chapter that Mr Cheng's bi-national sentiment and transnational longing are shared by compatriots in Hong Kong, bringing a new perspective to the 'myth of return' among international migrants. Hong Kong and Canada are stations between which strategic switching is practised by households within an extended social field and at distinctive stages in the life cycle. Among some migrants such time-space coordination is meticulously orchestrated.

Ocean Crossings and Re-crossings

Conventional wisdom situates return migration as an overlooked appendage to the historic emigration/immigration 'narrative' of departure, arrival, and assimilation. The weight of the assimilation narrative, especially in the United States, has tended to obscure the significance of the return trip home. In his examination of the *Round Trip to America*, Mark Wyman (1993: 4) has suggested that in the period of mass immigration from 1880–1930, 'Returned immigrants rejected America and, it seems, American scholars have rejected them'. While the population data are flawed and discontinuous, the best estimates suggest that during this half-century as many as a quarter to a third of arrivals to the United States re-crossed the Atlantic Ocean to return home. Significantly, rates were lower for the older migration sources in northwestern Europe, but much higher for newer national origins in southern and eastern Europe.

In the more recent, post-1945 period, return migration has continued even among northwest Europeans in such culturally compatible settings as Australia and Canada. Estimates suggest that as many as 20–30 per cent of Britons returned to the United Kingdom from these seemingly harmonious destinations (Hammerton 2004). For other groups, such as Turks in Germany, Italians and Greeks in Australia, or West Indians in Britain, return migration was a prospect long contemplated, for many ultimately a myth that was not enacted, but for others a transition prepared for by earlier return visits (Baldassar 2001; Duval 2004) and undertaken usually at retirement (Gmelch 1980; King 1986; Western 1992; Byron and Condon 1996; Thomas-Hope 1999).

A more recent repatriation process has been the appearance of the so-called 'brain exchange', complicating the earlier emphasis on the emigration and 'brain drain' of the global South's brightest and best to the countries of the global North. In developing countries like China and India an emergent high technology industry has led to return migration by citizens who had moved to western nations as students and young professionals, but who now see career and entrepreneurial opportunities in their countries of birth (Iredale et al. 2002). Return migration of the highly skilled has been encouraged by targeted programmes in some nations, notably China and Taiwan, including the construction of science parks as specific labour attractions for expatriates with high levels of human capital (Luo et al. 2002; Tsay 2002). Following reticence in the immediate shadow of Tiananmen Square (Zweig 1997), the option of return now appears more attractive for some among these contemporary Chinese emigrant 'first generations' of the 21st century.

But the tale of return migration has itself been complicated by current transnational developments. Return migration extends the linear model of the migration cycle to a circular model with an imputed re-adjustment and assimilation to the country of origin. The return has frequently been anticipated by earlier visits and by remittances that may well include funds for the construction of a new family house (Owusu 1998; Duval 2004). In this manner transnational connections are now recognized as important in facilitating return. Nonetheless return has an air of finality, of completing the circle of ocean crossings. But for some migrants return migration is less a final adjustment than another stage in a continuing itinerary with further movements ahead, whether unexpected or, as we shall illuminate, eagerly awaited.

To understand more fully the motives and implications of return migration in a transnational context, we organized seven focus groups in Hong Kong consisting of 56 returnees from Canada. They were identified both from personal contacts and from notices placed on the web sites of the alumni clubs associated with the University of Toronto and the University of British Columbia. For the latter institution, with its strong trans-Pacific linkages, the Hong Kong alumni club is the largest outside Canada. The observations of the Hong Kong sample are interspersed with ethnographic interviews conducted with economic migrants from Hong Kong still living in Vancouver.[2] This latter group are typically at older stages in the life cycle, and thereby cast light upon the space-time positioning of the returnees, who we argue are at a station now that may well not be permanent, but rather represents one point in a life-long trajectory of moves 'back and forth' across the Pacific Ocean.

2 These interviews were part of extended research among Hong Kong and Taiwanese migrants to Canada conducted since 1996. Ethnographic interviews have been conducted with some 200 households. See Ley (1995, 1999, 2003), Waters (2002, 2004) and Ley and Waters (2004).

Transnational Hong Kong

Transnationalism invokes a travel plan that is continuous not finite. Immigrants never quite arrive at their destination because they never quite leave home. Indeed the whole problematic of 'home' can become extraordinarily complex in an age with increasing levels of dual citizenship, labour contracts with short-term visas, family members located on opposite sides of national borders, and fast and ever cheaper lines of contact between nations. The life-world of the transnational migrant is stretched across space (Jackson et al. 2004); or, as one of our informants told us, the Hong Kong migrant would like to work in Hong Kong and sleep in Canada.

Much of the early transnational literature has been concerned with the relatively short-distance, and relatively inexpensive, movements between American cities and migrant origins in Central America and the Caribbean islands (Rouse 1991; Mountz and Wright 1996; Portes et al. 1999; Glick Schiller and Fouron 1999; Potter et al. 2005). In this research, in contrast, we are considering longer range movement, and a more costly trans-Pacific air journey of 12–13 hours between Hong Kong and Vancouver, the closest global city landfall on the American continent. Hong Kong, with its special and ephemeral constitutional status both in the past and in the present, is inherently transnational, 'not so much a place as a space in transit' (Abbas 1997: 4). From the mid-1980s, alarmed at geopolitical futures in East Asia, tens of thousands of middle-class residents left Hong Kong. In part they were enticed by welcoming immigration policy in Canada, Australia, New Zealand, Singapore, and to a lesser degree Britain and the United States. Canada in particular ran a pro-active immigration programme that recruited economic migrants from East Asia, and 380,000 migrants from Hong Kong arrived between 1980 and 2001, including 100,000 receiving visas through the business immigration streams, and another 64,000 securing entry as skilled workers. The numbers leaving the British colony and making the crossing to Canada reached a peak of over 44,000 in 1994 and for a decade Hong Kong was the leading immigrant source to Canada, and particularly its Pacific province of British Columbia.[3]

Fearful of closer ties with China in 1997, and precipitated by the Tiananmen Square massacre in 1989, many made the crossing for geopolitical reasons. We were told by one Hong Kong returnee that:

> I moved to Canada in 1989, when the Beijing massacre happened. But actually my parents already had the intention of moving to Canada to secure a better

3 For a selection of research on aspects of this migration, see Li (1992), Mitchell (1993, 1998), Skeldon (1994, 1995), Ley (1995, 2003), Wong and Ng (1998), Olds (1998, 2001), Wang and Lo (2000), Rose (2001), Wong (2003), Waters (2002, 2004) and Ley and Waters (2004).

> future for us. They were really concerned about Communist China and what that implied for Hong Kong in the future. Especially on my father's side, his family had experienced brutal treatment from the communist government because they were land owners.[4]

The 1989 massacre was the decisive trigger motivating migration for some households, but its impact added force to other motives that had already raised the issue of present insecurities in people's minds:

> ... because of the 4 June massacre, also because of the 1997 handover. That was the primary reason. The second reason was better education. We were all very young at that time, and my parents arranged for us to go to school there.

For others the educational motive was primary, in order to introduce one's children to the perceived superior (and more accessible) opportunities of Canadian schools and universities.

> It was more for our education. They [parents] think they have better opportunities over there. At that time it wasn't that easy to get into one of these universities in Hong Kong, so they thought it would be better for us.

For others again there was an emphatic 'quality of life' mandate, with appreciation of Canada's outdoor environment and available social and leisure services, information confirmed through transnational family networks:

> Because my aunt is there and my grandma was in Canada as well. So my father just wanted to live there. He loves Canada. The environment is very good, it's good for living.

But ominously (and ironically) interviews disclosed that economic factors did not appear to be prominent among the list of motives behind migration, although many new arrivals landed in Canada as economic migrants, including the largest single national group of millionaire immigrants granted admission as business investors and entrepreneurs.

Indeed, the government's business immigration programme has not unfolded as expected (Ley 2003). For a range of reasons, and despite their impressive pre-migration business experience, many Hong Kong migrants found economic success elusive in Canada. The business culture was far more regulated than they were familiar with, language was frequently a problem, and many who chose to invest in the ethnic enclave economy found cut-throat

---

4 These quotations come from returnees interviewed in focus groups in Hong Kong by the authors. Any names of respondents that appear in the text are fictitious.

competition in a saturated market. There is a suggestion too that a number were not fully committed to the task, but were seeking Canadian citizenship as an insurance policy, and once this had been secured they would return to an advantageous pre-existing economic niche in East Asia. Mak (1997) has noted that some Hong Kong firms gave favoured employees departing for Australia a two-year leave of absence, time to qualify for citizenship and then return to their former position. Senior managers at the Canadian Consulate in Hong Kong told us how in the early 1990s, they had confronted a new phenomenon for which their manuals gave them no answers. Well-qualified residents were applying for Canadian immigration visas, though the managers strongly suspected they had no real desire to live in Canada. Here were the 'reluctant exiles' whose anguished decision-making concerning departure was well-captured by Hong Kong social scientists at the time (Skeldon 1994, 1995).

Census data and tax returns reinforce the near unanimous view we heard from interviews on both sides of the Pacific about the limited economic success attainable in Canada.[5] Individual incomes for Hong Kong immigrants in Canada in 1996 were very low, with 45 percent earning less than Cdn$1000 a month, and mean incomes fell below half the level of returnees working in Hong Kong (DeVoretz et al. 2003). 'So I like Canada' observed one returnee, 'But the problem is I have to work there'. Another returnee weighed the alternatives: 'Everything is good in Canada except for job opportunities. The living standard is so good but the job opportunities [are] getting worse and worse'.

Not surprisingly there have been high levels of return migration. Exit data from Australia, (where unlike Canada such records are kept), suggest that as many as 30 per cent of 1990–1991 arrivals had returned to Hong Kong in short order (Kee and Skeldon 1994). Many households fragmented, with mothers and children left in Australia or Canada, while the father and husband assumed the role of 'astronaut', with his home and primary occupation in East Asia, undertaking long commutes for short visits on the 'Pacific shuttle' (Ong 1999) to see family members.

The Necessity of Return …

A weak job market and limited entrepreneurial opportunities were primary, but they were not the only factors prompting return. Some immigrants to Canada had treated their move single-mindedly as a means to gain a passport and thus neutralize their political anxieties in East Asia. With completion of a three-year residence requirement, they could return to Hong Kong in security, a Canadian passport in their pocket. 'If there are no political problems, it

5 There is considerable evidence of a comparable lack of economic success in Australia and New Zealand. See Ho and Bedford (1998), Ip, Wu and Inglis (1998), Burrill (2000) and Chiang (2004)

wouldn't matter which citizenship I have'. we were told. In addition to protection, Canadian citizenship also offered greater flexibility in visa-free international travel. So, 'Before it was insurance. Right now it's for the convenience'; and again, 'I strongly agree ... I find a [Canadian] passport is very, very convenient for me to go anywhere'.

The fact of return was already registering in Hong Kong media in the early 1990s, even before the peak year of emigration had occurred. In December 1993, an account in the *South China Morning Post* announced that 'Brain drain slows as managers return' (Ng 1993). Over the next five years the pace of such stories quickened. 'Immigrants flee Canada recession for rosy territory' (Anon. 1994); 'Hong Kong returnees on the increase' (Wallis 1994); 'Brain gain follows tremendous brain drain' (Batha 1996); 'Luxury (Vancouver) homes for sale as migrants return to SAR' (Lyons 1997); and 'Emigrants return home to better prospects' (Wong 1998). Simultaneously, estimates of the number of Hong Kong residents holding foreign passports steadily inflated in media stories, reaching the giddy figures of 500,000–700,000 out of a population of around 6.5 million by the mid 1990s. Canadian passport-holders were regarded as the largest single group with estimates ranging from a low figure of 150,000 to a maximum of 500,000 by the Canadian Chamber of Commerce in Hong Kong.

In 1999 the Government of Hong Kong sought to establish a profile of returnees through its General Household Survey of 22,300 households, conducted with a response rate of 92 per cent (Government of Hong Kong 2000). Returnees were defined as those who had returned to reside in Hong Kong after spending at least two years of the previous decade in another country. While the survey produced an estimate of 120,000 returnees, this figure was discounted as subject to 'substantial under-reporting ... [o]wing to the rather sensitive nature of the subject' (2000: 48). Nonetheless the profile of returnees was informative, and confirmed media reports that members of the departed economic elite were re-establishing residence in the territory. Returnees were three times as likely as the general population to be in the top income bracket of more than HK$30,000 a month. Almost three-quarters were employed as professionals, managers and administrators, and over half of the adults had university degrees, a rate five times greater than the population at large. There was an under-representation of children and the elderly, and a heavy over-representation of adults in their 20s and 30s. Thirty-five per cent had formerly lived in Canada, 24 per cent in Australia or New Zealand, 12 per cent in the United Kingdom, and 11 per cent in the United States.

More recently, special runs of the 2001 Census of Hong Kong for the entire population of 6.4 million have revealed a similar profile of returnees (DeVoretz et al. 2003). This analytical estimate is also a partial count, since the census includes Hong Kong residents in 2001, born in Hong Kong, who were living outside Hong Kong, Macao and China in 1996. The figure is an undercount not only because many emigrants from Hong Kong were born in China, but also because only returnees from the period 1996–2001 are caught in the census questions. No doubt too 'the rather sensitive nature of the subject' again encouraged under-reporting. Nonetheless this 2001 data are of great interest with some 86,000 returnees enumerated, 40 per cent of them moving from Canada, and with an equal share of men and women. The cohort was primarily in a career-building stage. The largest single group of Canadian returnees, 37.5 per cent, were young adults, aged 20–29, with another 21.5 per cent aged 30–39. Half the returnees from Canada had university degrees (70 per cent of these earned overseas) and the same proportion held professional or assistant professional positions. The elite nature of this returning cohort was rounded out by earnings levels that were two-thirds higher than the level of the overall resident population.

The consistency of the two data bases confers some confidence in identifying an ideal typical returnee as a well-educated professional, bi-lingual or better, in early career and with considerable earnings capacity. This description counters the typical retirement age profile of the returnee to North America, Europe and the Caribbean, but agrees with descriptions of skilled workers in Australia likely to repatriate to Taiwan or China (Guo and Iredale 2002). Our Hong Kong interviews and focus groups amply filled out this profile, and the deployment of human capital resources that it implies. Economic motives for return dominated all others. Comments such as these are characteristic:

> Promotions, opportunities, money. I think it's much better here. Here you work hard, but you get your promotions, your money.' You do work up the ladder. With a lot of my friends who graduated at the same time [they] are still in the same position [in Canada] or have only got one promotion, and it's been three or four years now …'

Or again:

> The only reason I want to come back is to find a job. Because in Canada it's not easy to find a job.

Likewise,

> I would say the working environment is better in Hong Kong. Like earning more money. Lower tax. That's the main issue I would say, lower tax. More opportunities here. I would say it's not hard to earn HK$15,000 a month for a fresh grad. But it would be super hard for the fresh grad to get a really good job in Vancouver or in Canada.

Finally,

> Q. *How much more do you think you get paid in Hong Kong?*
>
> Including [lower] tax, 300 per cent.

There is a sense of finality to these abstracted quotations that could easily be used to amplify the abundant newspaper stories of economic dissatisfaction in Canada leading to return migration and restitution of a territorial bourgeoisie. But the narratives of return and sojourning come to imply as much coherence and closure as the immigration-assimilation genre as we delved further. The focus groups revealed a much more complex and unfinished set of personal and family trajectories.

… At Least for the Medium Term

Recall that the government household survey and analysis from the 2001 Census both suggested a concentration of childless households in their 20s among returnees. Our focus groups reinforced the attractiveness of Hong Kong for this age cohort:

> [I'm here] because of job opportunities. Yes, mainly. And because I'm still young enough so I can tolerate the environment. And I'm more mobile I guess. Just to give it a shot here.

Hong Kong evokes terms like fast, bustle, energy, lively:

> We've all previously mentioned before, the working system, the energy, in this little place is actually much higher than we have in Canada.

And from the same focus group:

> Hong Kong is more bustling, it tends to be more lively, there's more of a night life … and it's more attractive to young people.

Respondents are working long hours to launch or consolidate their careers; focus groups were held of necessity over a working dinner in the evening as participants left a heavy business or professional day. But now we must intervene in this narrative of return and ask the question: to what extent is there an expiry date to this energetic pursuit of career development?

A number of families interviewed in Vancouver who had made the decision *not* to return alluded, sometimes emphatically, to the desire to escape a life in Hong Kong shaped by the single-minded pursuit of economic advancement (Ley and Waters 2004). In the words of one couple:

> *Mr Yee:* When we were in Hong Kong we both felt very busy for life, and we both wanted some life changes. And so we travelled a lot around the world to Australia, States, Canada. We went to Toronto and Vancouver. One morning in the summer in Vancouver, I stepped out the door of the

> house of my distant relative. I felt the air so fresh, and the sun so bright and everything so beautiful. And then I said to my wife, that's the place that we want to go …
>
> *Mrs Yee:* [In Hong Kong] it's the pressure you can hardly face because the whole society is so rushed, you know, and life is so busy that you can hardly slow down a little bit to enjoy life … It's both too busy for adults and the kids. So we want to slow down our pace a little bit so we came …

Mr and Mrs Yee were older than the young adults who formed the vanguard of the returnees to Hong Kong, and they had school-aged children. They identified the positive quality of life dimensions of Canadian society – slower paced, environmentally attractive, family friendly – qualities that were also acknowledged by our Hong Kong respondents, but had been set aside in deference to economic opportunities. But will there be a future date when these assets will be capitalized on by the returnees as well?

Some of the respondents, many single and in their 20s working hard to establish a career, projected themselves into the future to the status passage of having a family. They would then consider for themselves the decision made for them by their parents, and some would renew the trans-Pacific migration cycle. For others, in their 30s, the decision was at hand:

> I think my daughter will probably go back to Canada for her studies. Being that the education system here is such a mess. Going to international school [in Hong Kong], it costs so much money. She has to go back.'

Besides the passport, Canadian family networks remain, facilitating re-settlement. 'My family is still there. I have very close connections with them. So I call them probably, sometimes, at least 10 times per month.' Added to the presence of family and familiarity, the prospect of education and the quality of life, thought of Canada arouses generally positive memories, making the prospect of return plausible:

> I think we all have some special feeling for Canada. It's like a second home. I still have a brother who lives in Canada. Besides that, I like to eat some Canadian food, watch Canadian TV and all stuff like that. When I come back to Hong Kong for a couple of months I miss that and I want to go back.

In a transnational social field there is no finality to movement, but always the prospect of another 12-hour flight and another sojourn. Consider the following biography outlined by one of our focus group members, Simon, and his various trans-Pacific moves with their shifting motivations at different status passages through the life course.

> I immigrated in 1989. I really love Canada. Before we immigrated we would go to Canada for vacation, two to three times a year. The reason why I like Canada is because my younger brother studied there. After I've visited him there, I fell in love with Canada … He was in Vancouver. But my wife never wanted to go there. She really didn't want to immigrate to Canada, because she had a lot of friends in Hong Kong … After 4 June 1989, my wife was willing to immigrate to Canada.

Note, so far, frequent visits initiated by a brother who had been sent to Canada for education. This pattern of education-led migration is not at all uncommon (Waters 2004), though for Simon's family it took the political horror of Tiananmen Square to overcome his wife's diffidence. However, landing in Canada does not address the problem of economic well-being. Simon continues:

> We've been there for over 12 years. I've always been flying back and forth. I was an astronaut for about four or five years when we first moved to Vancouver … After that I decided to move to Vancouver permanently. So I sold my business in Hong Kong and moved to Vancouver permanently. But now I'm back here by myself. I'm working in Hong Kong, while my family members are staying in Vancouver.

The need for economic achievement meant that Simon's business in Hong Kong was maintained, and he adopted the identity of an astronaut, engaged in Ong's (1999) 'Pacific shuttle' between work in Hong Kong and his family on the Pacific coast of North America. Eventually this arrangement became socially unsustainable, the family business was sold, and he moved to Vancouver. But not permanently, for career objectives could not be sustained in his new home, and he is now back working in Hong Kong, once again an astronaut. This is not, however, the end of his frequent flyer miles, for another status passage is looming.

> Now, my son likes Canada, but he would like to come back to Hong Kong to work after his university education. He said that it's difficult to find a job in Canada. Actually, he prefers to live in Canada.
>
> Q. *So if your son comes back to Hong Kong, would you and your wife move here permanently?*
>
> Yes, we do have a plan. My wife and I have made a commitment that we would stay in Hong Kong for five years, and then we'll move back to Canada … Actually, I was already planning [in the 1980s] to save enough money in Hong Kong and go to Canada to retire.

We see from this remarkable family history that the two sides of the Pacific Ocean are a single social field transgressed seemingly at will at different stages of the life cycle in response to family needs that can be fulfilled more satisfactorily at one site or the other. In general terms it is economic activity that is the recurrent pull to Hong Kong, while quality of life attractions draw the family back repeatedly to Vancouver. In their own way, Simon's family have perfected a mobility path that optimizes each site in the social field for its own assets as they become relevant at discrete stages in the life cycle.

While few families have achieved this level of life-stage and movement synchronization, we see similar examples (and variants) of this over and over again in the biographies of our respondents. The enlargement of the social field means a dispersal of family members on both sides of the Pacific, a scattering of parents, siblings and children according to which station works best for their present stage in the life cycle. It means some confusion as to where exactly is home. It leads in Simon's case to him speaking as a resident in Hong Kong and telling us that 'We've been there [Vancouver] for 12 years'. A fusion of 'here' and 'there' has occurred in his mind because they are part of a single, if geographically diffuse, life-world.

And Then Retirement

We heard numerous examples of such careful synchronizations of space and time in an expanded social field. There was, for example, discussion about the best age to transfer children to schooling in Canada, the dominant view being around grade 10 when a sufficient level of Chinese language and culture had been absorbed, but in time for preparation for provincial examinations in Canada. This period also coincided with the much-feared Hong Kong Certificate of Education Examination, whose avoidance sometimes precipitated family migration in the first place (Waters 2004). But a prominent and unexpected topic for deliberation was the renewed transnational mobility anticipated at retirement.

Retirement is a significant status passage, frequently associated with migration (Rogers et al. 1992). For many it is a time to move away from metropolitan cores toward quieter settings with enhanced quality of life. For some it is a time to go home, including migrants whose life earnings have been secured and saved in metropolitan centres of the global north. The traditional/conventional view sees Greeks, Italians, Turks, and West Indians, amongst others, returning from diaspora to their homelands. It is here that the transnational longing of skilled Hong Kong returnees leads to a novel trajectory through a seamless social space that crosses oceans and national borders, passing from their native place of work to their adopted place of rest. Interviews with a small sample of skilled Taiwanese Australians planning repatriation to Taiwan suggested the same spatial strategy of a double return, first to Taiwan to work, but with the prospect of a later return to Australia upon retirement, if not sooner; 'many saw returning as yet another temporary move and anticipated retiring or returning regularly to Australia' (Guo and Iredale 2002: 35).

Simon, still working in Hong Kong as we heard above, was already thinking in the 1980s of moving to Canada upon retirement, just as Albert Cheng, with whom this paper began, has retirement plans in Canada even at the peak of his media and political career in Hong Kong. A surprisingly large number, perhaps half, of the respondents in our focus groups had the same forward planning in mind. Here then is a particularly transnational double-take on the myth of return. There is a well-recognized expression among this transnational population 'Hong Kong for making money, Vancouver for quality of life'. At retirement the balance of these two valuations undergoes a significant reassessment.

> There's polluted air, polluted water. Almost everything is polluted in Hong Kong so when I retire I don't want to stay in Hong Kong. There's no fun, you can't go to fishing, you can't go skiing. If I can, I want to go back there tomorrow. But I can't afford to go back there right now because I need to make a living.

This theme was pervasive.

> I'd say my parents really love Vancouver and Canada. Weather, clean air, environment, the living style and standard, they love all sorts of things. I think they would choose Vancouver after they retire.

And again:

> I will consider moving back after retirement, though I still have thirty years to go. My dream is to go back to Vancouver for retirement … [My parents] plan to be there after retirement. My dad will retire in seven years. He will live there with my mum, because there it's more comfortable.

For some families, retirement offers the prospect of emotional re-integration across fragmented spaces:

> Q. *Did your parents move to Canada with you?*
>
> No, my mum was with us, my dad stayed in Hong Kong.
>
> Q. *Like an astronaut?*
>
> Yes, for eleven years.

> Q. *He has not moved there so far? Has he been an immigrant?*
>
> No ... not until he retires.

Hong Kong in contrast is regarded as too expensive, too crowded and too polluted for comfortable retirement.

Within a patriarchal family structure there is some evidence of variability between men and women in prioritizing the economic/Hong Kong *versus* the lifestyle/Vancouver stations in the social field. Economic gradients typically impel men more strongly. In several of the interviews quoted above, it is the astronaut men (like Simon) who are in East Asia, while their wife and family are in Canada. A returnee told us how economic under-performance in Canada gnaws at male self-esteem:

> When you are not able to find a job, or earn enough money or work at your former position, especially for a male, they'll feel that they are useless. Some people who are used to being a boss, after they went to Canada they had to distribute newspapers or to work as a driver for a living. As their social class lowered dramatically, they also suffered serious psychological depression. I think, other than money this is another important reason why many have returned to Hong Kong.

It is the men who are typically the initiators of return to Hong Kong. In contrast, after a difficult settling in period, women seem to adjust well to the opportunities for self-development in Canada (Waters 2002). In one of our focus groups a couple had very different views about the desirability of return, differences that led to some emotional disagreement, with the husband acknowledging 'She didn't want to come back to Hong Kong. So we're planning to go back there when we retire'. The husband of another returnee family had found a job that required him to travel to Beijing during the week, returning only at weekends. His wife was left isolated and nostalgic:

> I was there with my daughter facing the four walls. I was depressed ... I was there in Hong Kong all by myself. I have friends there but they're busy too, as you know well ... Hong Kong people! I've been so used to having different activities in Vancouver, like playing tennis, choir singing and pot luck dinners. I don't seem to be fitting into the Hong Kong lifestyle any more.

The age when retirement will occur depends in good measure upon the scale and speed of acquisition of mobile capital. The low incomes of recent Hong Kong migrants in Canada, reported earlier, are to some extent the result of underemployment associated with early retirement, particularly for those in the business immigration streams who have significant international assets but a limited Canadian cash flow (Ley 2003). As significant pre-migration business experience is a prerequisite of entry through these streams, household heads are typically in their forties when they land in Canada. Within a few years many will be considering retirement. For example, Mr Liang, a senior manager in a multinational corporation in Hong Kong, entered Canada through the investor stream of the business programme, an immigration track that permits passive investment without active entrepreneurialism (Ley 1999). Upon landing in Vancouver, at age 52, he began his retirement, identifying through his space-time decision-making the appropriate station in an expanded social field for a man of his age and economic achievement:

> ... you can't earn any money here. If you have enough money you can come. Just stay here and relax ... You cannot expect to have a [comfortable] life, or to earn good money here because the tax here is so high. So you must have earned enough to come ... I would never think of doing business here. Actually most of my friends, they retire, because they are all my similar age. All my good friends are retired, I mean those who come recently, we are all retired people.

Hong Kong is for making money and Vancouver is for quality of life. Back in Hong Kong one of our focus group members, less than 15 years behind Mr Liang, is considering the same trajectory.

> I'm almost 40. I'm almost thinking of retirement. I have to plan for my kids' education. They might end up somewhere else on the globe. I don't care where they want their education. For myself, I really want to retire in a place I feel really comfortable in ... I think I will have to end up in Canada.

It is important to consider some of the potential consequences, if and when spatial mobility at retirement is activated as planned. We can anticipate the return to Canada of a population with considerable savings to spend in real estate and consumer goods and services. It is worth reminding ourselves that expansionary effects upon the consumption sector were registered in the Vancouver regional economy during the peak years of Hong Kong immigration during the 1986–1996 decade, while the real estate market fell back appreciably with the population exodus beginning in the mid-1990s (Lyons 1997; Walkom 1997). Such expansionary pressures should be expected to happen again as this still-mobile cohort returns on retirement. But in addition to these private sector impacts, there could well be public service costs associated with the arrival of this significant retirement cohort with appreciable assets but modest taxable income. They are also likely to locate disproportionately in the Vancouver region where the quality of life is high, where a large, institutionally complete, diasporic community exists, and where the cross-Pacific air journey to and from Hong Kong is the shortest among major North American centres. Awareness of these implications was registered by some respondents.

> I know I'll definitely go back to retire because I'm Canadian so I can have all those benefits that retired people get. So I know I'll go back for retirement. Prior to retirement I may not, depending on financial, social, there's a lot of different things. But definitely for retirement!

Conclusion

The methodological advantages of conducting this transnational research at two sites, in Hong Kong and Vancouver, were frequently evident. The capacity to interview students in Vancouver, graduate career builders in Hong Kong, middle-aged parents with school children and retirees back in Vancouver, together rounded out their transnational experiences and strategies across the life cycle. Importantly, we could more fully understand the routines of everyday life at each site which made that location favourable now, but perhaps less so in the past or the future, as we heard migrants project themselves back and forward in space and time. As they lived across two territories linked by dense electronic messaging and frequent travel, stretching relationships and resources across space, so our field research needed to range between the two (or sometimes more) sites that comprised their blended social field.

Return migration is not a sufficient description of the hyper-mobility of transnational citizens living, presently, in Hong Kong, and for whom the two sides of the Pacific are part of a single life-world. As we have seen, their continuing itinerary over time undercuts conventional accounts that portray return migration as circular migration with its own logic of arrival, assimilation and closure. Instead there is a perennial openness to further movement at distinctive passages in the life cycle.

There are other theoretical issues at stake as well, such as the creation of a social space that transcends (and challenges) national borders (Ong 1999; Ley 2003), the meaning of citizenship and identity in a fluid bi-national residential history (Kobayashi 2004), the elasticity of family relations stretched across the Pacific Ocean (Waters 2002), and, as Albert Cheng expresses so courageously, the portability of political (and other) values between transnational sites. These issues merge readily with policy questions specific to different national territories, when, for example, the jurisdiction collecting career stage taxes, is not the same jurisdiction that dispenses retirement stage benefits. A larger calculus is required here that systematizes costs and benefits more fully than has been possible to date (for example, by Wang and Lo 2000).

Meanwhile at different stages in the life cycle the migrant capitalises upon one or other site in this trans-Pacific life-world, passing from one station to another, but always with openness to the option of return. As one focus group respondent reminded us, that process can go on even beyond retirement.

> I had no choice but to come back [to Hong Kong] to make a living. ... But I still wish to live there (Vancouver) after retirement, because I like the lifestyle there. But I prefer dying in Hong Kong, not there.

Acknowledgements

We are grateful for the skilled field assistance of Guida Man and Priscilla Wei, and to George Lin for his generous logistical support in Hong Kong. The study was funded by grants from the Strategic Research Programme of the Social Sciences and Humanities Research Council of Canada and from the Vancouver Centre (RIIM) of the Metropolis Project. This chapter first appeared as an article with the same title in *Global Networks* (2005) 5(2): 111–27. The permission of *Global Networks* to include this research article by David Ley and Audrey Kobayashi in this collection is gratefully acknowledged.

References

Abbas, A. (1997), *Hong Kong: Culture and the Politics of Disappearance*, Hong Kong: Hong Kong University Press.

Anon. (1994), 'Immigrants flee Canada recession for rosy territory', *South China Morning Post*, 10 April.

Baldassar, L. (2001), *Visits Home: Migration Experiences Between Italy and Australia*, Melbourne: Melbourne University Press.

Batha, E. (1996), 'Brain gain follows tremendous brain drain', *South China Morning Post*, 6 November.

Burrill, R. (2000), 'The Business Skills programme: is it delivering?', *People and Place*, 8 (4): 36–42.

Byron, M. and Condon, S. (1996), 'A comparative study of Caribbean return migration from Britain and France: towards a context-dependent explanation', *Transactions of the Institute of British Geographers*, 21: 91–104.

Chiang, N. (2004), 'The dynamics of self-employment and ethnic business ownership among Taiwanese in Australia', *International Migration*, 42: 153–73.

Devoretz, D., Ma, J. and Zhang, K. (2003), 'Triangular human capital flows: empirical evidence from Hong Kong and Canada', in Reitz, J. (ed.), *Host Societies and the Reception of Immigrants*, La Jolla CA: Center for Comparative Immigration Studies, University of California, San Diego, 469–92.

Duval, D. (2004), 'Linking return visits and return migration among Commonwealth Eastern Caribbean migrants in Toronto', *Global Networks*, 4: 51–68.

Glick Schiller, N. and Fouron, G. (1999), 'Terrains of blood and nation: Haitian transnational social fields', *Ethnic and Racial Studies*, 22: 340–61.

Gmelch, G. (1980), 'Return migration', *Annual Review of Anthropology*, 9: 135–59.

Government of Hong Kong (2000), *Returnees to Hong Kong*, Census and Statistics Department, Special Topics Report No. 25.

Guo, F. and Iredale, R. (2002), 'The view from Australia', in Iredale R, Guo, F. and Rozario, S. (eds), *Return Skilled and Business Migration and Social Transformation*, Wollongong: Centre for Asia Pacific Social Transformation Studies, University of Wollongong, 21–37.

Hammerton, J. (2004), 'The quest for family and the mobility of modernity in narratives of postwar British emigration', *Global Networks*, 4: 271–84.

Ho, E. and Bedford, R. (1998), 'The Asian crisis and migrant entrepreneurs in New Zealand', *New Zealand Population Review*, 28: 71–101.

Ip, D., Wu, C.-T. and Inglis, C. (1998), 'Settlement experiences of Taiwanese immigrants in Australia', *Asian Studies Review*, 22: 79–97.

Iredale, R., Guo, F. and Rozario, S. (2002), 'Introduction', in Iredale, R., Guo, F. and Rozario, S. (eds), *Return Skilled and Business Migration and Social Transformation*, Wollongong: Centre for Asia Pacific Social Transformation Studies, University of Wollongong, 1–19.

Jackson, P., Crang, P. and Dwyer, C. (2004), *Transnational Spaces*, London: Routledge.

King, R. (1986), *Return Migration and Regional Economic Problems*, London: Croom Helm.

Kobayashi, A. (2004), 'Transnationalisms: developing theoretical and policy understandings', paper presented to the Ninth International Metropolis Conference, Geneva, September.

Ley, D. (1995), 'Between Europe and Asia: the case of the missing sequoias', *Ecumene*, 2: 187–212.

Ley, D. (1999), 'Myths and meanings of immigration and the metropolis', *The Canadian Geographer*, 43: 2–19.

Ley, D. (2003), 'Seeking *homo economicus:* The Canadian state and the strange story of the Business Immigration Program', *Annals of the Association of American Geographers*, 93: 426–41.

Ley, D. and Waters, J. (2004), 'Transnationalism and the geographical imperative', in Jackson P., Crang, P. and Dwyer, C. (eds), *Transnational Spaces*, London: Routledge, 104–21.

Li, P. (1992), 'Ethnic enterprise in transition: Chinese business in Richmond, BC', *Canadian Ethnic Studies*, 24: 120–38.

Luo, K., Guo, F. and Huang, P. (2002), 'China: Government policies and emerging trends of reversal of the brain drain', in Iredale, R., Guo, F.

and Rozario, S. (eds), *Return Skilled and Business Migration and Social Transformation*, Wollongong: Centre for Asia Pacific Social Transformation Studies, University of Wollongong, 71–90.

Lyons, D. (1997), 'Luxury homes for sale as migrants return to SAR', *South China Morning Post*, 20 August.

Mak, A. (1997), 'Skilled Hong Kong immigrants' intention to repatriate', *Asia and Pacific Migration Journal*, 6: 169–84.

Mitchell, K. (1993), 'Multiculturalism or the united colors of capitalism', *Antipode*, 25: 263–94.

Mitchell, K. (1998), 'Reworking democracy: contemporary immigration and community politics in Vancouver's Chinatown', *Political Geography*, 17: 729–50.

Mountz, A. and Wright, R. (1996), 'Daily life in the transnational migrant community of San Augustin, Oaxaca and Poughkeepsie, New York', *Diaspora*, 6: 403–28.

Ng, L. (1993), 'Brain drain slows as managers return', *South China Morning Post*, 24 December.

Olds, K. (1998), 'Globalization and urban change: tales from Vancouver via Hong Kong', *Urban Geography*, 19: 360–85.

Olds, K. (2001), *Globalization and Urban Change: Capital, Culture, and Pacific Rim Mega-projects*, Oxford: Oxford University Press.

Ong, A. (1999), *Flexible Citizenship*, Durham, NC: Duke University Press.

Owusu, T. (1998), 'To buy or not to buy: determinants of home ownership among Ghanaian immigrants in Toronto', *The Canadian Geographer*, 42: 40–52.

Portes, A., Guarnizo, L. and Landolt, P. (1999), 'The study of transnationalism: pitfalls and promise of an emerging research field', *Ethnic and Racial Studies*, 22: 217–37.

Potter, R.B, Conway D. and Phillips J. (eds) (2005), *The Experience of Return: Caribbean Perspectives*, Aldershot and Burlington, VT: Ashgate.

Rogers, A., Frey, W., Rees, P., Speare, A. and Warnes, A. (1992), *Elderly Migration and Population Redistribution: A Comparative Study*, London: Belhaven.

Rose, J. (2001), 'Contexts of interpretation: assessing urban immigrant reception in Richmond, BC', *The Canadian Geographer*, 45: 474–93.

Rouse, R. (1991), 'Mexican migration and the social space of postmodernism', *Diaspora,* 1, 8–23.

Skeldon, R. (1994), *Reluctant Exiles? Migration from Hong Kong and the New Overseas Chinese*, Armonk, NY: M.E. Sharpe.

Skeldon, R. (1995), *Emigration from Hong Kong: Tendencies and Impacts*, Hong Kong: Chinese University Press.

Thomas-Hope, E. (1999), 'Return migration to Jamaica and its development potential', *International Migration*, 37: 183–203.

Tsay, C.-L. (2002), 'Taiwan: significance, characteristics and policies on return skilled migration', in Iredale, R., Guo, F. and Rozario, S. (eds), *Return Skilled and Business Migration and Social Transformation*, Wollongong: Centre for Asia Pacific Social Transformation Studies, University of Wollongong, 91–112.

Walkom, T. (1997), 'Hong Kong exodus worries BC', *The Toronto Star*, 29 September.

Wallis, K. (1994), 'HK returnees on the increase', *South China Morning Post*, 13 July.

Wang, S. and Lo, L. (2000), 'Economic impacts of immigrants in the Toronto CMA: A tax-benefit analysis', *Journal of International Migration and Immigration*, 1: 273–304.

Waters, J. (2002), 'Flexible families? Astronaut households and the experiences of lone mothers in Vancouver, British Columbia', *Social and Cultural Geography*, 3: 117–34.

Waters, J. (2004), 'Geographies of cultural capital: international education, circular migration and family strategies between Canada and Hong Kong', unpublished dissertation, Department of Geography, University of British Columbia, Vancouver.

Western, J. (1992), A *Passage to England: Barbadian Londoners Speak of Home*, Minneapolis: University of Minnesota Press.

Wong, K.-W. (1998), 'Emigrants return home to better prospects', *South China Morning Post*, 29 September.

Wong, L. (2003), 'Chinese business migration to Australia, Canada and the United States: state policy and the global immigration marketplace', *Asia and Pacific Migration Journal*, 12: 301–36.

Wong, L. and Ng, M. (1998), 'Chinese immigrant entrepreneurs in Vancouver: a case study of ethnic business development', *Canadian Ethnic Studies*, 30: 64–85.

Wyman, M. (1993), *Round Trip to America*, Ithaca, NY: Cornell University Press.

York, G. (2004), 'Canadian opts for Hong Kong vote', *Globe and Mail*, 5 August.

Zweig, D. (1997), 'To return or not to return? Politics *vs.* economics in China's brain drain', *Studies in Comparative International Development*, 32: 92–125.

Chapter 8
Bittersweet Home? Return Migration and Health Work in Polynesia

John Connell

In Pacific island states, the migration of skilled health workers is not new, yet few studies trace any facet of this emigration and circulation process, and relatively few examine other components of skilled migration in the region (Liki 2001; Voigt-Graf 2003; Voigt-Graf et al. 2007). Even less is known about the return migration experiences of such skilled workers. To begin filling this gap, this chapter examines the return migration of skilled health workers, mainly nurses, in two of the smallest Pacific island states, Niue and the Cook Islands.

Rapidly increasing international migration of skilled workers has been perceived as a response to the accelerated globalization of the service sector (Iredale 2001; Findlay and Stewart 2002; Lowell 2002; Xiang 2007). Professional services such as health care have been central to this new internationalization of labour, as demand for skilled health workers in developed countries emerged early, generating the first literature on 'brain drain' (Gish and Godfrey 1979; Mejia et al. 1979). Reduced recruitment of health workers in developed countries followed declining birth rates, and more diverse employment opportunities for women, many of which offer superior wages and working conditions, and greater prestige and respect. Jobs in the health sector were increasingly seen within many metropolitan states as too challenging and poorly paid, while demand has also increased with the aging of developed country populations, and higher expectations of medical care (Connell 2008b).

Ironically, these are broadly the same reasons that skilled health workers in Pacific island states have left their own national health services. Most islanders have migrated for financial reasons, but such exigencies are combined with social reasons (including further education for themselves and other family members overseas) and further qualified by professional concerns over the inadequacy of the local health sector in terms of the broad conditions of employment (Connell 2004). For small Pacific island states, the migration of skilled workers has been perceived as something of a one-way process, a loss of scarce human capital, a critical brain- and skill-drain, and a major problem hindering island development. Moreover, training of skilled health professionals is particularly costly because of its relatively long duration, the high costs of teaching materials and techniques and the very limited resources of most small island states. Significant return migration would alter this scenario, or so it is believed.

Return Migration

Globally, there is remarkably little information on return migration for most migration source countries, hence it remains 'the great unwritten chapter in the history of migration' (King 2000: 7). It is a 'compound and complex process', often over an extended period of time, that involves 'flows of goods and capital as well as the ideas, attitudes, and skills of the migrants themselves' (Thomas-Hope 1985: 157, 172). Recent studies, especially in the Caribbean, have recognized the diversity of return migration, noting the multiplicity in reasons for return (Thomas-Hope 1999; Gmelch 1992; Conway et al. 2005), but coverage elsewhere is minimal. Return migration in the Pacific has been largely overlooked for example, despite half a century of international migration and return in most Polynesian island states. Where return has been examined it commonly has been characterized as a migration movement dominated by retirees and those who have failed elsewhere (Maron and Connell 2008). Those who remained overseas were therefore 'the successful', and though many publicly expressed intentions of return, in practice they embraced extra-regional permanence (Macpherson 1985; Alexeyeff 2004).

This is at least a far cry from the nineteenth century when returning Cook Islanders were reported by missionaries to have 'an unsettling influence on island life and formed a 'depraved and vicious' element advocating such sins as prostitution and the consumption of fermented liquors' (Gill 1946, cited by Curson 1973: 107). Overall these early return migrants were regarded as least likely to make a significant positive contribution to their extended families and their home territories.

More recent studies have suggested a rather different situation exists nowadays. For many Cook Islands migrants, for example, the acquisition of new skills overseas was a contributing factor in the decision to return, particularly with the ensuing rise in social status and income. Most returned for social reasons: to support families, because of homesickness or to construct a house. Although many took a significant time to get a job, education overseas nonetheless proved highly beneficial in acquiring good jobs at home. Out of a small sample of returnees, as many as half the men and three quarters of the women moved out of the public sector into the private sector, setting up small businesses on return. Most expressed satisfaction following their return but were concerned about such issues as transport, housing and salaries (Hooker and Varcoe 1999; Rallu 1997; see also Marcus 1981: 60). In Tonga, returnees represented a cross section in terms of age and employment, including unskilled workers, skilled health workers and those with second degrees (Maron and Connell 2008). Yet in both these two national contexts there were relatively few such skilled return migrants, compared with the number of those who had left.

The return migration of skilled health workers is assumed to be relatively limited in most places, though data are scarce, hence the benefits from enhanced overseas skills – that would constitute a compensatory brain gain, and which continue to be touted as a significant gain from skilled migration (Kirk 2007) – may be few. If significant return migration of skilled health workers occurs, return migrants with their (usually) enhanced skills and experience would constitute a positive transfer of human capital. If migrants return, even after a relatively short time overseas, with new enthusiasm and perhaps also capital, there may be major gains from such migration.

Return Migration of Skilled Health Workers

Fragmented evidence from many parts of the world suggests that return migration of skilled health workers largely fails to occur for the same reason that migration previously occurred. Migrants are unlikely to be tempted back by a system that they left, at least in part because of its perceived shortcomings. They may return for reasons that have nothing to do with rejoining the public health system, however, but rather because of 'bonding' or family reasons. What research suggests, although it is not comprehensive in coverage by any means, is that few (or even no) migrant health workers wish to return to their home countries, with most preferring to move onwards, through a hierarchy that culminates in the United States (Troy et al. 2007; Percot 2005; Ball 2008), while prior intentions of return are quickly diluted after departure (Connell 2008b). Indeed, the incidence of return migration of skilled health workers in contemporary times has been perceived to be so slight that Kingma (2006) has referred to it as a 'myth'; the 'myth of return' of émigré nurses, who are expected to go back, but do not. In several contexts even the limited return of skilled health workers has remarkably little to do with the health sector's attractiveness or career prospects. In Kerala, and more generally in India, and in the Philippines, nurses either do not intend to return, or return to take up positions of perceived higher status outside the health sector (Percot 2005; Thomas 2007; Ball 2008). Jamaican nurses overwhelmingly returned for family reasons, and most returned to work in the health sector; however the more qualified nurses, with the most to contribute, rejected this because of poor salary and working conditions (Brown 1997: 206–9). While some returning health workers might take up higher status positions in the health system, most are simply lost to health care.

Overall therefore, the limited evidence on the return migration of skilled health workers elsewhere in the world suggests that relatively few return. Those who do return to the local health sector tend to have been overseas only briefly (or were bonded to return), while those who returned after longer periods were less likely to return to the (public) health sector but rather favour employment commensurate to their skills and experience elsewhere. In other words, those who returned to local health sectors had been away a short time, and had gained relatively little in the way of new skills or experience, while those who had been away longer and gained valuable skills were unlikely to add to the human capital stocks of the local health sectors of their ancestral homelands (Connell 2008b, 2009b).

The Heart of Polynesia: The Cook Islands and Niue

The Cook Islands and Niue are two of the smallest island states in the Pacific. Both were former colonies of New Zealand, neither has achieved full independence or membership of the United Nations and their residents are New Zealand citizens. Like most other Polynesian states (notably the rather larger Samoa and Tonga, and the even smaller Tokelau) they are characterized by emigration and the majority of ethnic Niueans and Cook Islanders live overseas. In 2006 the population of the Cook Islands was about 14,000 (but there were about 58,000 people of Cook Islands ancestry in New Zealand) while the population of Niue was about 1500, and 22,500 people of Niuean descent lived in New Zealand. Limited land and natural resources, isolation and fragmentation (in the Cook Islands), weak infrastructures and governance all pose problems for administration and development, and economic growth has been very weak in recent years. Both countries experience a poverty of opportunity, and are highly dependent on overseas aid. Migration has consequently increased, mainly to such metropolitan states as Australia, New Zealand and the United States, to the extent that the two territories have declining populations – the only significant instances of 'absolute depopulation' in the Pacific. Both Niue and the Cook Islands have been classified as classic MIRAB (MIgration, Remittances, Aid and Bureaucracy) states; 'rentier states' where the external sector dominates the domestic sector (Bertram and Watters 1985) and migration is of critical economic importance.

For both states, metropolitan countries have also traditionally been the destinations for tertiary studies, but some doctors are educated within the region (at the Fiji School of Medicine). Nurses and other skilled health workers have usually been educated in Fiji or New Zealand. The health systems of both countries have been significantly affected by migration, particularly of doctors and more specialized occupations such as lab technicians and dentists. In both countries the emigration of doctors is considered to be more significant than that of nurses in terms of proportions who had migrated, their impact on the health care system and the costs of replacement (Connell 2005, 2009a).

A 'culture of migration' exists where migration is pervasive, based on historical precedent, part of everyday experience, perceived as legitimate, neither rupture nor discontinuity in personal and household experience, but an integral part of life (Connell 2008a). Migration is normative and mobility, intermittent return visits and return migration are integral elements to the mix. International migration has long had a critical and virtually uncontested role, centred on strategies for extended family/household development, rather than the outcome of individual decisions. A strong social component influences decisions concerning such contemporary overseas migration with the location of close kin and perceptions of family obligations being major influences. Like other migrants, health workers remain part of extended 'transnational corporations of kin' (Marcus 1981), whereby their migration is encouraged (or, at the very least, not discouraged) by the financial needs of family members who remain in the islands. In both societies, the migration and return migration of skilled workers is embedded in this broad context of community and continuity.

Migration of Health Workers in the Pacific

This chapter is part of a wider study of the migration of health workers in the Pacific region, undertaken between 2000 and 2005 (Connell 2004, 2009a), and focuses on the two smallest island states, and the health workers in those two states who were return migrants, alongside a few who had moved away. Overall 41 people (15 in Niue and 26 in the Cook Islands) were return migrants; two thirds of these were nurses, and relatively few were neither nurses nor doctors. Virtually all the nurses were women, and about two thirds of the doctors were men, a reflection of the 'traditional' gendered structure of the health sector in the two states.[1]

Migration of skilled health workers from Niue and the Cook Islands is primarily related to education (training overseas), economic issues (the professional incomes of skilled workers overseas), superior quality of life involving the employment context (better working conditions, facilities, opportunities for research and career development), and various social factors (educational opportunities for children, morale). In the Cook Islands especially this has been exacerbated by economic restructuring, reductions in the size of the public service, and deterioration in local working conditions, especially in more remote areas. Skilled health workers, like other professionals in small island states, feel they are isolated from trends in their profession and in the wider world and are conscious that they may miss out on new skills that will enable professional development (and, perhaps, future migration). Limited access to contemporary technology and training facilities were (and still are)

1 All quotations are from health workers in the Niue and the Cook Islands unless otherwise stated. The names have been changed. I am indebted to Hazel Easthope for her assistance with interviews in the Cook Islands. The title of the chapter is taken from a book of the same title describing the ambiguities and uncertainties of migration for Indo-Fijians in the central Pacific.

regular concerns, and telemedicine is yet to remedy this. Furthermore, in this first decade of the 21st century, local wages and salaries are even more ubiquitously seen as inadequate. Two thirds of all nurses and almost half (46 per cent) of all doctors are primarily motivated to migrate for income reasons (Connell 2004, 2008c). The importance of income to these health professionals, whether absolute or relative, can not be underestimated.

Becoming a Health Worker

Almost all had first gone overseas for tertiary education and training which occasionally led to prolonged stays. For some, their migration is only incidentally related to an employment or career goal, or more specifically, employment in the country of origin. Rather, migration decision-making has much more to do with attempts to improve the long term welfare and status of families. In both states many entered the health professions less out of altruism, or a particular interest in medicine, but through the recognition or 'hope' that this might be a means of maximizing or at least improving family incomes and welfare, and because scholarships were available. Parents encouraged their children to enter the profession for the same reasons. Employment in the health system thus enables migration as much as it is an instigator of it.

A predilection for migration occurs even before taking up employment or training. Various reasons were given why people had originally joined the health service, despite some initial lack of interest on their part or limited knowledge of what it might entail. Many nurses were interested 'since childhood' but the specific reasons ranged from 'I always wanted to be a nurse; I admired nurses in their white uniforms' to 'My father was a surgeon and my brother a doctor; I was always interested and challenged by the medical field but really it's in the family'. Parents were sometimes discouraging, partly because nursing was often seen as 'dirty work' – a contrast to the glamour of the uniforms – or had 'too many responsibilities' for the limited income.

Some were encouraged by parents to join the health sector, sometimes because of the assured government employment and reasonable income and sometimes to look after their own health needs; one nurse who wanted to be a teacher noted; 'My parents were selfish – only thinking about themselves – for me to look after them'. Others became nurses because they saw this as their responsibility: 'I wanted to help people and earn some money for my family'. Many workers thus joined the health sector out of their desire to look after and support their extended family, though at least as often senior family members encouraged support.

One 30-year-old Cook Islands nurse, one of the few who had never been overseas, indicated

> I was chosen to be a nurse so that I could stay here and look after the land and my parents while my other brothers and sisters went overseas [but] I

would like to go overseas to get more experience in nursing, to work in a more modernized situation because I fear I may lose what I have learned if I stay here, but it's hard to get my family to understand since they chose me to stay and the rest of the family are overseas.

Families were thus making decisions not only about jobs but at least implicitly about the future residence of their children. That entering a health profession enabled, even ensured, overseas migration, was implicit for most workers from Niue and the Cook islands, unlike larger Pacific states where it was more deliberately spelled out (Connell 2009a, 2009b). Nonetheless, some did emphasize that it provided an opportunity for migration: 'I became a dentist basically to get out of the country', and kin supported these choices because jobs were prestigious and remittances would follow.

Availability of scholarships was a major influence on choice of employment. In both countries, but especially Niue, scholarships were partly a function of national need hence high school students jumped at the opportunity of any scholarship, even if the health profession was not necessarily their own priority. This was true for all groups of health care occupations, and the linkage between jobs and scholarships was even closer where more specialized skills were required, as for example in dentistry, or where students had limited knowledge of what was required, such as being a lab technician. While many had gained scholarships in their preferred fields, at least as many had not. Even getting their first choice was not necessarily straightforward: 'I got a scholarship because the Minister of Health at the time was my mother's cousin.' Choice was constrained. 'I wanted to teach but I got a place in nursing school', 'I wanted to be a travel agent, and my parents thought nursing was a dirty job, but it was all that was available.' Not surprisingly the outcomes were not ideal: 'I'm still confused as to whether I want to be a nurse or a teacher', and students recurrently dropped out of courses. Some later recognized they had made a mistake; six of the 10 Niuean nurses for whom adequate data were available would have preferred some other form of employment at the time they began their nursing careers.

Especially in Niue, because of the limited number of local positions, some returnees found that their training was not directly relevant to the tasks that they were being asked to do. Although some enjoyed the training programmes, and later the jobs, several workers were relatively unsatisfied in their particular positions, and experienced low morale. Somewhat earlier Heyn (2003: 35) similarly found in a more wide-ranging study that 'many people on the island expressed a desire to have a different job'. It is very difficult for all workers to find the most appropriate and satisfying jobs in such small territories, it seems.

Consequently the health workforce does not necessarily emerge from those who know most about it, are most committed to its goals and necessarily best suited to it. The means of selection in these small states effectively condemn some people to a job with an unacceptable level of dissatisfaction that only makes matters worse. Nonetheless, becoming a health worker in such constrained circumstances also shows how education attainment represents a critical investment in human capital for many islanders, who then may be able to acquire employment in a wider range of contexts. Hence, there is such a demand for tertiary education that young people will grab scholarships in whatever profession(s) they are offered, regardless of their personal goals and proclivities.

The Return of Health Workers to Niue and the Cook Islands

With reference to the specific geographical context of this chapter's focus, if not common in other Pacific island territories or elsewhere in the global South (as mentioned earlier), a significant number of health workers have returned to the two Pacific island states of Niue and the Cook Islands. Indeed out of the overall sample of 51 health workers interviewed in the wider study (Connell 2004, 2009a) some 41 (80 per cent) had returned from overseas. Significantly only six of these (all in the Cook Islands) had returned after employment overseas; fully 35 had returned after training and had never been employed outside the home country. Not all returnees, who were previously health workers, work in the health sector (but because the survey data came from workers in the health sector this proportion cannot be identified). Almost all the health sector work-force were therefore return migrants, but those who had been employed overseas were conspicuous by their absence.

Only exceptionally was return migration linked to the work context, in terms of being promoted (one nurse) and finding it easy to get a job (one nurse), as a rationale for return. Most nurses stated outright or implied that their 'homeland' bonds were important reasons for them to return, demonstrating that the bonding system was an effective way of encouraging the return of those training overseas. One said that returning was good 'for her people and country' but not for her and her family, believing that she could have done better for her family if she had remained overseas. Another said that had she been single she did not think she would have returned to the Cook Islands.

Most therefore returned because they were bonded to do so, often despite some degree of reluctance. But, as New Zealand citizens, Niueans and Cook Islanders are able to move freely between their home islands, New Zealand and Australia, and most individuals have as many (or more) relatives overseas as at home. Hence, return migration for whatever reason is never necessarily the end of the story. Beyond bonding, social reasons influenced return, for both doctors and nurses, including the rather nebulous but constant 'it's my home country', to being with friends and relatives and accompanying a spouse home. Employment in the health sector was not an incentive to return, and for a significant proportion return was a duty, or obligation, as much as an act of free will. Migrants tended to return at key moments in the life course – after training or before marriage, for example, or when children had graduated – but at least as often when their parents experienced particular needs. Return migration was often of households or individuals who had returned to care for aging parents. In other words, many returned not necessarily at times of their own choosing or of their own volition.

Experiences 'Back Home'

On balance, return was a positive experience since in both states the majority wanted to remain rather than re-return again. But, most were well aware that their attitudes might change in the future, and many remained dissatisfied with return (in varying degrees) and were dissatisfied with their specific roles in the workplace (and the survey could not capture those who had dropped out of the health sector on or before return). Some of those who had returned to the health sector were dissatisfied enough to be seeking other employment and some had found other jobs. Many expressed a wish to get into private business or simply 'be a part-time shopkeeper in addition to nursing to earn some extra dollars', or 'set up my own used clothing business'. Some had achieved that sort of complementarity: 'the main benefit from returning is that I can also work in agriculture and now export beans, capsicum, zucchini and pawpaw' (this, from a 61 year old pharmacist). Not only had health workers taken up such jobs and left the health service, but others who had returned to the health sector wanted to move into something more economically rewarding.

Economic benefits, writ large, from return migration could best be gained by working outside the health sector. This gave returnees greater income and a degree of individualism, independence and autonomy, and resulted in many return migrants being part of multi-income households. Returning health workers were particularly likely to establish a business on their return, having accumulated enough savings for this to be possible; a pattern that occurs more widely amongst Polynesian returnees in larger states (Brown and Connell 1993; Maron and Connell 2008). Almost all Niueans, and many of the Cook Islanders who had returned, lived in households where more than one other person was employed, invariably in the public sector (subsidised by New Zealand), hence though they might not have returned for income reasons they had significant household incomes (Connell 2007).

Those who wanted to leave again invariably sought higher wages or superior access to technology and training. 'I want more experience with serious cases and surgical patients', 'Doctors are treated better overseas; here there is corruption and politics, and I'm losing contact with developments in medicine.' Returnees in this group were usually relatively young, unlike the 52-year old nurse, who admitted: 'it's too much to change again at my age'. What may bring repetitive migration to an end is that eventually it becomes 'too hard to start again at the bottom', alongside a concentration of friends and relatives in one place or the other.

More influential than the attractions of overseas salaries, technology, living standards and lifestyle was discontentment in the workforce, fuelled by comparisons with the situation overseas. Underpinning tensions that existed at work were local salaries and conditions that compared unfavourably with the health sectors that they had returned from and/or been trained in. Most frustration related to work conditions and lack or recognition of skills and knowledge which sometimes amounted to at least perceptions of blocked promotion. Those who had been forced to start again at the bottom of the employment hierarchy were particularly frustrated by wages, conditions and the hierarchical rigidity. Others were frustrated by the lack of adequate technology. Some resented having had to give up specialized work to return as generalists, and there was almost unanimous disapproval of the level of wages and the minimal overtime payments; concerns which echo throughout the region (Connell 2009a). The 'lack of equipment' and 'stress, pressure, too much work' all affected morale. Yet, and despite these strongly-held criticisms, many returnees still expressed satisfaction at working for the government, because of the prestige, reliability and stability that it provided.

Smallness poses particular problems: 'people have no idea when doctors may be resting and call in or phone up at all hours'. Small staff numbers offer little diversity in terms of medical skills, hence referrals tend to be numerous, while those with particular professional specialties may effectively be almost always on call. Ironically, pressures were often greatest in islands at a distance from the centre, as information and equipment reached regional centres slowly and staffing levels were often high. 'I hate the long hours, and the phone calls at night when I want to sleep, but it's all part of the job.' Only the Cook Islands has regional-care centres and it is no accident that only the southern island group, relatively close to the capital, have Cook Islands doctors (though turnover is very high) whereas the remote northern group have immigrant Burmese doctors.

In considerable contrast, somewhat unusually, and unlike most parts of the Pacific, 'boredom' and low job satisfaction was brought about by modest-to-minimal workloads. Many health workers, and not merely those with specialized training, did not have enough work to occupy them. This was generally true of most Niuean public sector employment; so that in government offices 'island-time' hangs most heavily. In the health system this was particularly so at night and resulted in workers dwelling on other frustrations of the workplace, and feeling that they were not usefully contributing. This also inhibited the development of a positive 'work ethic'. In some more specialized areas there was rarely enough work to uphold professional standards. While this enabled workers to be somewhat flexible over their actual hours at work, there is a limit to flexibility and it does not compensate for reduced motivation or the challenges to maintain standards:

> I like being here but I will probably end up going overseas- I want to put into practice what I have learned in Fiji – I need more hands-on experience – we just get bored – there are no emergencies and emergency medicine was exciting.

Frustrations were particularly directed at hierarchical management structures.

Many returnees stressed nepotism and favouritism in island health care systems, alongside the annoyance of being unable to implement changes that work elsewhere: there's 'a lot of favouritism; people get high salaries and top positions with less experience'. Some of those who had returned 'find that promised jobs are given to someone else when we return' since 'it's not what you know, but who you know'. More generally there were:

> Bad attitudes; people were unsupportive of what I came back with, and my new knowledge, however much I wanted to share what I had got from my new degree, which simply made we want to move away again. I only have one more year of bonding and then I can go back to New Zealand. (38-year-old nursing tutor)

Problems of leadership, or its absence, and the lack of support for junior staff can easily become critical issues in a small workplace with a hierarchical structure. Typical complaints were these: 'Bosses are a big problem; they make life miserable; they don't help; they discourage you more than assist; it's much better when they are not around', and often 'there is too much gossip around – sometimes you simply have to get up and walk away'.

Therefore, a familiar concern of returnees was that the extra skills and qualifications they had enthusiastically, laboriously and dutifully gained overseas actually counted for very little. Perhaps the greatest frustration for many was their inability to use these skills and experience. 'Domineering bosses' prevent and discourage innovation and change. This was well summarized by a forty year old nurse in the Cook Islands:

> It's hard to make changes here. I try to make changes but the other nurses pull you back. They don't want to change. They think you're bigheaded.

This kind of 'crab antics', that discourages innovation and ideas, is not unusual in many workplaces in Pacific island states, but there is a particular irony here since almost all Cook Islands and Niue nurses have at least trained and often worked overseas themselves. Older workers who have worked for many years, have certain privileges and profess themselves 'too old' to move – yet are resentful of newcomers who seek to make changes that may challenge their particular role. Stability in a small workforce with limited turnover discourages change. Established personnel often had a wealth of knowledge and experience and while some innovations and changes might have failed, some or all of the benefits of overseas training and employment are lost. Initial enthusiasm wears off and it becomes just another job, much less a vocation, to be tolerated rather than enjoyed. The anticipated potential benefits from return migration were very hard to deliver, since attitudes to return migrants were both complex and contrary: often welcomed in theory but spurned in practice.

* * *

A Salutary Vignette: Jane Tupoku's Story – Niue, New Zealand, Saudi Arabia, Australia and Canada as her Transnational Social Field

In some circumstances, qualifications and skills acquired overseas were simply too specialized, notably in Niue, where a more general multi-skilling characterizes the small, local health workforce, and complex procedures are undertaken through overseas referrals. Jane Tupoku's 'story' exemplified this situation.

Jane grew up in Niue but completed her final high school year in New Zealand. She then trained there as a nurse, since several of her relatives were nurses and she perceived it to be a career that would earn a good living and help her family. After graduation she returned to Niue for two years to fulfill her scholarship bond obligations. She then returned to New Zealand for a year's midwifery training before again coming back to Niue. At first she enjoyed being back, there were many responsibilities and her New Zealand education meant that she was expected to take on several activities.

> After a bit, I no longer had enough patients to retain my interest; there were too many minor mundane things. I used to wonder what they were doing in New Zealand since I was curious about new techniques and ideas. There was not enough interaction here – too few people. You have to get used to sitting. Now I can't work here – my work is much too specialized. In New Zealand I got used to the fast-moving lifestyle – people are over-relaxed and over-confident here. They just take their time and don't seem to care but do it when they feel like it. It's not much of a challenge – I like to be exposed and on my toes, more than just relaxing I want to learn about new technologies. In New Zealand when you need help it's there.

After again fulfilling her bond, Jane went back to New Zealand to work in Wellington, disappointing her parents though she did not intend to go permanently. In New Zealand she enjoyed the lifestyle and people, liked the type and variety of nursing she was doing, interaction with other staff, access to superior resources and the privacy of personal life. Two years later she went to Jeddah in Saudi Arabia, after being contracted by an Auckland agency:

'other nurses had been there and had both positive and negative experiences, the money was great and it was an adventure though we were naïve about it all'. She stayed there seven years, enjoying the work (where she worked in a specialized neonatal unit), the community life, and the excellent facilities, despite certain cultural constraints. Briefly she worked alongside other Niuean nurses. Eventually she learned to teach scuba diving all around Egypt, Sinai and elsewhere. After leaving Saudi Arabia she returned to Niue for three months, but then went to Cairns in northern Australia where she could again combine scuba diving with work in a specialized neonatal intensive care unit. There was plenty of work, including work with the Flying Doctor service delivering premature babies, but she eventually found this too quiet and wanted to do a Masters degree.

Through a magazine advertisement Jane found that Vancouver was recruiting neonatal nurses and went there since she had heard positive things about the health care system, there was a bigger unit there and she could a further degree. Two years later she returned to Niue in 2004 in the wake of Cyclone Heta that had devastated Niue and destroyed the hospital, and undertook voluntary shifts in the makeshift hospital to relieve the staff. Cyclone Heta made her question why she was in Vancouver, so far from home and family, though she maintained a close connection with family, telephoning a couple of times a week to her parents, emailing here brothers and sisters just as frequently, and returning from Canada and Saudi Arabia at least once a year. Nonetheless in 2005 she gained permanent residence in Canada, where she enjoyed the seasons, the 'healthy lifestyle and mentality'. She had never married. By 2006 when she had again returned to Niue for her father's funeral, she felt that there would only be one move after Canada and that would be back to Niue. But, as she put it;

> I can't work here, my nursing is much too specialized, and I'm not inspired here – I would have to get used to sitting around. There is no big picture here, about patients or community, no element of compassion, nursing is just a chore, there is no leadership and inadequate equipment to do a great deal. Eventually the good ones become despondent and leave.

Not long after the funeral, she returned to Canada once more.

Jane exemplifies those migrants who have undoubtedly succeeded overseas, and although retaining various close contacts at home, find it extremely difficult to return despite family connections and social obligations. Some careers and expatriate social lives are more demanding and fascinating and simply do not exist at home for such 'high fliers'. Return would demand too many social and economic sacrifices, constrained opportunities and, because Jane had siblings who remained at home, there was less pressure on her to return.

* * *

While many were optimistic about their own ability to make contributions to change and development at home, through their training and experience overseas, they were less optimistic of their ability to make positive changes than return migrants were in the larger workforces of Fiji, Samoa and Tonga (Connell 2009b), and more critical of the hierarchical, oppressive structures in these smaller islands. Indeed there was not always enthusiasm for the efforts of returnees. As one local Cook Islands doctor stated: 'returning migrant nurses are too old and out of touch, but we will take anything: we have no choice'.

Nonetheless, a fair proportion of returnees enjoyed their work in health, had few grievances with the system, and found a considerable degree of fulfilment. 'I wanted to pay back the Niuean people and reduce the language barrier that is associated with expatriates, but we understand local values and conditions' and 'I like being able to work with and help my own people'. The job itself offered satisfactions: 'I like caring for very sick patients and seeing them get well and going home to their families – but hate it when they go down', or, quite simply, 'I like this job; no other job would be me'. In some circumstances however similar altruism may better explain why health workers took up their positions initially, rather than remained there, as such feelings do not necessarily sustain a career. 'If I had not been married I would not have come back but now I am here I want to serve my country. The Ministry of Health has invested a lot in me and so I should work from them. Besides both me and my husband have good jobs and this is where our relatives and friends are.' Duty accompanies desire, in such cases.

The health service lends itself to certain forms of satisfaction: 'meeting people from different islands' and 'working closely with the community and getting respect from people' or 'seeing the emergence of new life'. Teamwork enthused many. Others felt they could use their overseas knowledge and much satisfaction was implicit in simply doing a good job and knowing one was contributing. Inevitably frustrations were also attached to 'failures' that were implicit in the very nature of practicing medicine: just as many people were concerned about their inability to reduce pain, prevent death or stimulate good practices as had frustrations in the workforce.

In the upper echelons of the health sector, returning health professionals – head nurses, doctors, surgery technicians, anesthetists, etc. – often valued the particular freedoms and privileges of high status positions – such as their ability to travel frequently to overseas conferences and workshops. Working in the public service creates a geographical paradox: 'I stay here because there are opportunities for travel overseas. If you want to travel you have to work in Niue … I try to go [to Auckland] for meetings, to catch up with friends, kin and KFC.' In six years one health worker had been to conferences or training workshops in nine different countries. Remaining in the Pacific actually enable these returnees' mobility. Particularly in Niue there was perhaps something to be said for being a 'big fish in a small pool' – 'Here I am king of my island;

everywhere else someone is bossing you'. Power had its perks; 'I was going to return overseas but now I'm in charge.'

Many of the pleasures of return were detached from the workplace. 'Lifestyle' – which usually meant family and friends, alongside having one's own house and often land – made sometimes difficult or poorly paid working conditions much more tolerable. Owning land 'back home' offers security and stability. A typical comment was:

> I love the lifestyle here – the freedom and safety, and quietness. It's not safe in New Zealand. I own my own house and land. It's a slower pace of life but I did have some culture shock when I got back [after three years in Auckland] and had to teach myself to speak Cook Islands Maori again.

For many there was a 'more comfortable pace of life' where it was 'more relaxed – you can do your own thing'. Peace, tranquillity and safety were also valued: 'it's more free, and there's less stress for bringing up children' whereas 'I was scared of going out in New Zealand'. In contrast to their considerable criticisms of the workplace, many of those who had moved back emphasized the pleasant climate, the more relaxed pace of life, or simply the familiarity of the home country and its people, as immensely satisfying.

Circulation

Even some of those who were aware they were doing a valuable job, and enjoyed it, found it difficult to balance this with their knowledge of 'higher wages and a better quality of life overseas' that represented a constant lure and temptation. In the existing climate of overseas recruitment and substantial demand for skilled health workers, it was usually older workers who did not see this as both a temptation and a possibility. Nonetheless the majority of those who had returned and were working in the health sector wished to stay for a variety of reasons, usually related to their age, the location of close relatives, the employment status and aspirations of their partner (usually their husband), ownership of land, a house and sometimes a small business. Those who wished to return again, to New Zealand or Australia, focused on a combination of superior incomes, training opportunities for themselves and a better life for their children.

Those most likely to contemplate, and achieve further migration, are the relatively young, who are more likely to be discontented with various aspects of island life, and who can perceive a career ahead. A study of Niuean students in New Zealand who had chosen not to return concluded that the principal reasons for this included limited opportunities, low salaries and a generation gap where young graduates perceived themselves as innovative but saw older people as being stuck in their ways and who would give no support to new ideas. Other factors included no job challenges, all the family being resident in New Zealand, lack of a clear future – including the ability to use acquired skills, lack of mentoring and support, no obvious career path and a feeling that Niue was not going forward (Brunt and Morris-Tafatu 2002: 54).

Among those who wished to stay and those who sought to go – and there was no clear-cut distinction between them – many exemplified uncertainty and ambivalence in their decision-making. Commonly, there was recognition that certain imponderables, notably changing family needs, would influence whatever their individual preference might be. Thus a Fijian nurse working in the Cook Islands stated;

> I would love to go next to Australia where I can get extra training and develop my career to become a theatre nurse. My children will really benefit from that and my husband [a chef] could take more cooking courses. However my parents want me to be home in Fiji to look after them so I shall probably have to do that.

Relatively few saw their present location and position as permanent and most at least contemplated some kind of future change.

Many had already migrated more than once as part of a cycle of circular migration, in response to different social and economic factors. The repeated mobility of Tevai, a Cook Islands nurse, almost entirely in response to extended family needs (Hooker and Varcoe 1999: 94), is a classic example of such a circular migration scenario.

* * *

Repeated Circular Migration: Tevai's Story

Tevai, born on the island of Aitutaki, lived there, in Tahiti and then New Zealand before finishing high school in Australia where she went on to train as a nurse. Tevai had been working in Sydney as a nurse for six months when her father died and she was called back to her family in Aitutaki. For the next few years she adopted a dual lifestyle, living alternately in Aitutaki and Australia. In Australia, Tevai earned good wages nursing and was able to send part of her wages home to her mother. When her mother became ill, Tevai found it necessary to spend more and more time in Aitutaki. In 1991 she applied for a full-time nursing job at the hospital in Rarotonga. Although she considered the wage to be very poor compared with wages received in Australia, Tevai wanted to live closer to her mother. After her mother's 60th birthday, the family had a meeting and decided that Tevai had done her share of looking after her mother and that Tevai's sister would come back from Australia to take over. Tevai wants to go back to Australia in a couple of years to earn more money but intends to return to Rarotonga eventually and build a house on the land given to her by her mother (Hooker and Varcoe 1999: 95)

At virtually every stage in life, therefore, Tevai and her family balanced the opportunities for income generation in Australia, to provide economic support, with the need to be close to family in the Pacific to provide social support. Variants of Tevai's story, and of the transnational corporation of kin, are common: 'I went to Australia for two years and worked in a nursing home but came back when I was needed to look after my uncle's five children' (nurse's aide, Cook Islands). Social obligations underpin these and other migration histories.

* * *

Island health workers are thoroughly familiar with the costs and benefits of both mobility and stability. Many have experienced both phases, often several times. Migration is seen as more individualistic, whereas those who stay are said to be committed to community (usually expressed as village or church) and nation. Those who leave are perceived as unwilling to put in the time needed to maintain such local social ties. Migration and stability can both be seen in a positive and negative sense, with benefits and costs, respectively.

'Stayers' remain because the benefits presently outweigh the costs, but all are aware that this is most likely a function of the particular time in their life-course that relates to the ages of children, the health, age and location of parents and the particular position of themselves (and their partners, and sometimes children) in the workforce (Connell 2008a). 'I'm stuck here now because of my grandmother – we keep an eye on her – but the rest of my family has gone and we will join them.' By contrast for older Niueans: 'My family is here building up our homes and our families. I don't want to have to go and start again. The lifestyle here is good.' Choices over migration hinge on sometimes quite small changes in these contexts, and are facilitated by the ease of migration to extra-regional destinations- New Zealand and Australia, in particular.

About a third of the nurses and half the doctors wanted to migrate again 'soon', though such intentions might be quite different from action. This new, or next, phase of migration would be to acquire better education and/or provide new experiences for the individuals and for their children, underpinned by higher wages and a better job. But overall an economic rationale was always a significant factor. By contrast those who intended to remain wished to stay because it was home and where their relatives and friends lived. Returning and staying tend to be social phenomena, leaving an economic one, whilst mobility is a constantly unfinished story.

Conclusion: Return to the Islands

An exceptionally well established structure and culture of migration has become fully embedded in both island societies. Skilled health workers have moved to take advantage of superior wages and salaries, training and research opportunities and working conditions for themselves, to better the lifestyles of their children in terms of access to education, and to joining kin overseas. The 'skill-drain' is likely to continue, especially where there have been structural reforms that reduce public sector employment, wages and salaries remain unequal, working conditions are difficult and many kin are overseas. Most people have more kin overseas than at home. Policies that redress such circumstances have proved difficult for small island states, and island governments have rarely sought to intervene in the process of international migration. A significant 'brain drain' exists, evident in the health sector but equally evident in other contexts, from IT workers to teachers and footballers (Connell 2006), as national populations decline.

Return migration is not uncommon but, without bonding, the loss of health workers would be much greater. At no time during the past quarter of a century has there been substantial return, despite a pervasive ideology of return. Most return simply follows the completion of training overseas. Nor is return without problems; returnees inevitably compare more lavish facilities, and wages in the metropolitan states with those in their homelands. The factors that influence return migration include the climate, safety or the relaxed pace of life, or simply the familiarity of the home country and the presence of kin. Return is invariably little to do with economics or employment, but is for family reasons and not therefore necessarily at times of their own choosing (Maron and Connell 2008). In such circumstances, their successful adaptation on return is the more remarkable.

Migrants only rarely returned because of any great desire, other than duty, to work in the health sector, and employment in health was seen as inadequate in itself. Many who returned did so because they were also able to invest in other activities, usually in developing some form of business (often with their spouse), that would compensate for salary reductions. While many returned to work in the public sector, it is probable that at least as many moved into the private sector on return, because they had acquired capital and because of the extra autonomy and opportunities that this provided. Economic benefits, such as the ability to open a store, which may have been funded from remittances (Liki 2001; Brown and Connell 2004), lay outside the health care system, which often became for many mere supplementary employment (Brown and Connell 2006; Connell 2009a). Others returned because their spouse wished to return, or because it was jointly a valuable economic and social strategy.

The principal benefit from return migration should be the transfer of skills acquired overseas, especially within a familiar cultural context. Generally there is limited evidence that this occurs, as technology often differs and resources are absent (Kingma 2006: 201). Both countries lacked adequate absorptive capacity to receive many return migrants, and especially to benefit from the new skills and ability to work with new technology, that their return might entail. Significant constraints to using new skills, and to innovation generally, also rest in rigid hierarchies and uncomprehending and uninterested superiors in the local workplaces. Such problems were most evident in Niue simply because the workforce was so small that promotion prospects were both limited and predictable, while change was threatening. Return migrants may be 'agents of change' but conformity is usually more appreciated than change in these small island societies.

> More commonly, it is return migration that slowly changes islands but, wherever and however it occurs, and especially on the smallest islands, migration and change incite resentment, envy, tension and new perceptions of identity. (Connell and King 1999: 18)

Return results in some degree of confusion and uncertainty about identity, enhanced by the expectations placed on returnees, by individuals and by social institutions, their own recognition that they had changed and their inability to meet others' expectations.

Many migrants find return difficult, facing lower wages and standards of living, difficulty in establishing businesses, and simply culture shock. Their success and their return were sometimes resented, both in the workplace (where they sought to make changes and introduce new ideas etc) and in local society. Consequently returnees within the health sector were not particularly committed to contributing to the sector. They were more likely to resent the nepotism they found, that was quite different from the meritocracy of metropolitan states, and which hindered promotion and innovation, and so stayed for a relatively short period of time.

Yet, ironically, there is a need for more return migration. Both states have a shortage of health workers and both health services, at some cost, have employed migrants from elsewhere, especially as doctors (in the Cook Islands from Burma, Russia, Bulgaria, Fiji and the Solomon Islands, and in Niue from Samoa and New Zealand) but also as nurses, at least in the case of the Cook Islands where there are several nurses from Fiji. Despite return migration there is an overall loss of skills. Nonetheless return migration occurs, is reasonably balanced, and certainly not merely of failures and retirees, though some may have dropped out of the workforce, as 'failures' (but the methodology precluded their being recognized). However some of the most successful in the modern international world, such as Jane Tupoku, are least likely to return. The return migration of those with skills has thus tended to be limited, in part because those skills cannot necessarily be practiced locally (Conway et al. 2005; Connell 2009a), but more frequently because return migrants are poorly recompensed. Yet even the most established overseas migrants, such as Jane Tupoku, are rarely entirely divorced from home. Return is always possible, if unlikely.

The 'skill drain' belatedly provides some real gains through human capital transfers (with return migration), previous remittances and the investments of returnees (Connell and Brown 2004). For the Cook Islands, qualified and experienced people have returned and been able to use their skills in a range of occupations, but not primarily in public service (Hooker and Varcoe 1999: 96), though both the Cook Islands and Niue have public sector wages and salaries that have some comparability with other sectors in these destinations. Returnees are generally more educated than those who have stayed, tend to achieve social mobility on return by acquiring businesses or taking up management positions while also moving from the public sector to the private sector (Rallu 1997; Maron and Connell 2008). Consequently individuals and households have gained from migration, rather more than the wider society and economy.

Despite a considerable 'skill-drain', return migration is significantly greater than might have been expected where a general 'myth of return' has been suggested to prevail (Kingma 2006), and where almost all migration from the South Pacific island states has been of emigrant settlers rather than temporary contract workers. While numbers may be small, their impact is significant for both social and economic development, in terms of gains to health services (through new skills and wider experience) and to the economy for their (partial) investment of overseas generated incomes. Yet many are dissatisfied, and their return migration might be, in reality, just another episode or circuit in their unfinished, transnational migration cycle. Return migration, particularly of skilled health workers, is therefore more problematic than for most other returnees. Ultimately *ambivalence* is at the core of the Pacific 'culture of migration', where skills must be acquired overseas but dependent families remain at home. This is no less true of skilled migrants, as Jane Tupoku's story indicates, and emphasizes how migrants are seemingly forever caught between two worlds, how their contributions to national development are thus reduced, and whose journeys are never complete (Chapman 1991; Small 1997; Maron and Connell 2008); however much they may sometimes seem to be.

References

Alexeyeff, K (2004), 'Love food: exchange and sustenance in the Cook Islands diaspora', *Australian Journal of Anthropology*, 15: 68–79.

Ball, R. (2008), 'Globalised labour markets and the trade of Filipino nurses', in Connell, J. (ed.), *The International Migration of Health Workers*, New York and London: Routledge, 30–45.

Bertram, G. and Watters, R. (1985), 'The MIRAB economy in South Pacific microstates', *Pacific Viewpoint*, 26: 497–520.

Brown, D. (1997), 'Workforce losses and return migration to the Caribbean: A case study of Jamaican nurses', in Pessar, P. (ed.), *Caribbean Circuits. New Directions in the Study of Caribbean Migration*, New York: Center for Migration Studies, 197–223.

Brown, R. and Connell, J. (1993), 'The global flea-market: migration, remittances and the informal economy in Tonga', *Development and Change*, 24: 611–47.

Brown, R. and Connell, J. (2004), 'The migration of doctors and nurses from South Pacific island nations', *Social Science and Medicine*, 58: 2193–210.

Brown, R. and Connell, J. (2006), 'Occupation-specific analysis of migration and remittance behaviour: Pacific island nurses in Australia and New Zealand', *Asia Pacific Viewpoint*, 47: 133–48.

Brunt, A. and Morris-Tafatu, C. (2002), *Review of NZODA Scholarships and Training in Niue*, Niue.

Chapman, M. (1991), 'Pacific island movement and socioeconomic change: metaphors of misunderstanding', *Population and Development Review*, 17: 263–92.

Connell, J. (2004), 'The migration of skilled health professionals: from the Pacific islands to the World', *Asian and Pacific Migration Journal*, 13: 55–177.

Connell, J. (2005), 'A nation in decline? Migration and emigration from the Cook Islands', *Asian and Pacific Migration Journal*, 14: 327–50.

Connell, J. (2006), 'Migration, dependency and inequality in the Pacific: old wine in bigger bottles?', in Firth, S. (ed.), *Globalization and Governance in the Pacific Islands*, Canberra: Pandanus, 59–106.

Connell, J. (2007), 'At the end of the world: holding on to health workers in Niue', *Asian and Pacific Migration Journal*, 16: 179–98.

Connell, J. (2008a), 'Niue: embracing a culture of migration', *Journal of Ethnic and Migration Studies*, 34.

Connell, J. (2008b), 'Towards a global health "care" system', in Connell, J. (ed.), *The International Migration of Health Workers*, New York and London: Routledge, 1–29.

Connell, J. (2009a), *The Global Health Care Chain: From the Pacific to the World,* New York: Routledge.

Connell, J. (2009b), '"I never wanted to come home": skilled health workers in the South Pacific', in Lee, H. and Francis, S. (eds), *Transnationalism and the Pacific Islands*, Canberra: Australian National University E Press.

Connell, J. and Brown, R. (2004), 'The remittances of migrant Samoan and Tongan nurses', *Human Resources for Health*, 2(2): 1–21.

Connell, J. and King, R. (1999), 'Island migration in a changing world', in Connell, J. and King, R. (eds), *Small Worlds, Global Lives: Islands and Migration*, London: Pinter, 1–26.

Conway, D., Potter, R.B. and Phillips, J. (2005), 'The experience of return: Caribbean return migrants', in Potter, R.B., Conway, D. and Phillips, J. (eds), *The Experience of Return Migration: Caribbean Perspectives*, Aldershot and Burlington, VT: Ashgate, 1–25.

Curson, P. (1973), 'Birth, death and migration: elements of a population change in Rarotonga, 1890–1926', *New Zealand Geographer*, 29: 103–19.

Findlay, A. and Stewart, E. (2002), 'Skilled labour migration from developing countries', ILO International Migration Papers No. 55, Geneva.

Gish, O. and Godfrey, M. (1979), 'A reappraisal of the 'brain drain' with special reference to the medical profession', *Social Science and Medicine*, 13C: 1–11.

Gmelch, G. (1992), *Double Passage: The Lives of Caribbean Migrants Abroad and Back Home*, Ann Arbor, MI: University of Michigan Press.

Heyn, J. (2003), 'Migration and development in Niue Island', unpublished MSc thesis, University of Montana.

Hooker, K. and Varcoe, J. (1999), 'Migration and the Cook Islands', in Overton, J. and Scheyvens, R. (eds), *Strategies for Sustainable Development*, Sydney: University of New South Wales Press, 91–9.

King, R. (2000), 'Generalizations from the history of return migration', in Ghosh, B. (ed.), *Return Migration: Journey of Hope or Despair*, Geneva: International Organization of Migration, 7–56.

Kingma, M. (2006), *Nurses on the Move. Migration and the Global Health Care Economy*, Ithaca, NY: Cornell University Press.

Kirk, H. (2007), 'Towards a global nursing workforce: the "brain circulation"', *Nursing Management*, 13(10): 26–30.

Liki, A. (2001), 'Moving and rootedness: the paradox of the brain drain among Samoan professionals', *Asia Pacific Population Journal*, 16(1): 67–84.

Macpherson, C. (1985), 'Public and private views of home: will Western Samoan migrants return?', *Pacific Viewpoint*, 26: 242–62.

Marcus, G.E. (1981), 'Power on the extreme periphery: the perspective of Tongan elites in the modern world system', *Pacific Viewpoint*, 22: 48–64.

Maron, N. and Connell, J. (2008), 'Back to Nukunuku: employment, identity and return migration in Tonga', *Asia Pacific Viewpoint*, 49(2): 168–84.

Mejia, A., Pizurski, H. and Royston, E. (1979), *Physician and Nurse Migration*, Geneva: WHO.

Rallu, J.-L. (1997), *Population, Migration, Développement dans le Pacifique Sud*, Paris: UNESCO.

Small, C. (1997), *Voyages: From Tongan Villages to American Suburbs*, Ithaca, NY: Cornell University Press.

Voigt-Graf, C. (2003), 'Fijian teachers on the move: Causes, implications and policies', *Asia Pacific Viewpoint*, 44: 163–74.

Voigt-Graf, C., Iredale, R. and Khoo, S. (2007), 'Teaching at home or overseas: teacher migration from Fiji and the Cook Islands', *Asian and Pacific Migration Journal*, 16: 199–224.

Xiang, B. (2007), *Global 'Body Shopping': An Indian Labor System in the Information Technology Industry*, Princeton, NY: Princeton University Press.

Chapter 9
Returning Youthful Trinidadian Migrants: Prolonged Sojourners' Transnational Experiences

Dennis Conway, Robert B. Potter and Godfrey St Bernard

The twin Republic of Trinidad and Tobago has a resident population of approximately 1 million (2006) and a significant diaspora of émigré communities abroad that is estimated to exceed 20 per cent of its resident population. A national survey conducted in 1998 estimated there were as many as 10,000 return migrants who had returned to the twin-island country (St Bernard 2005). As well as the expected counter-stream of more elderly first-generation returning retirees, a significant proportion of this 10,000 is a self-selective cohort of young – and relatively youthful – 'Trini' transnational migrants of working age, who have decided to give it a try 'back home'. Some are second-generation returning nationals, or 'citizens by descent', having been born abroad. Others – the one-and-a-half (1.5) generation – left Trinidad and Tobago as children and were brought up abroad in the UK, Europe, USA and Canada, before returning home.

Another group had left Trinidad as school leavers in their late teens and early 20s for further education, skill acquisition and overseas experience, but stayed abroad after their studies and lived overseas for many years (between 12 and 30 plus, is our characterization), before eventually returning. We refer to this cohort of youthful pre-retirement returnees as prolonged sojourners, in large part because many viewed their stay overseas as temporary and had every intention of returning home sometime – sooner or later.

In this chapter, we focus on the transnationalism and return experiences of 24 prolonged sojourners, who were still relatively young and in mid-career and mid-family formation ages – in their 30s and 40s – when they returned. Our original sample of informants numbered 40; comprising nine second generation, seven one-and-a-half generation and 24 prolonged sojourners. Since most, or all, of the prolonged sojourners are living transnational lives, we are interested in their transnational practices, experiences and how these influence their adaptations to circumstances back home. We are interested in their reasons for going away and then returning, as well as their adaptation experiences on return. We are interested in their future plans, and whether they are staying for good or contemplate leaving and re-migrating again, or re-returning to the metropolitan society that had been their former home.

As a corollary of these specific questions, we are particularly interested in pursuing the suggestion that a relatively small number of these return migrants might effectively mobilize human and social capital and contribute directly to local development imperatives. Returning in their 30s and 40s, these returnees are prolonged sojourners rather than permanent emigrants, so their connections to their birthplace and homeland might be expected to be influential with regard to their 'development potential' and their willingness to return home to help 'make a difference', and 'give something back' (also see Potter and Conway 2008).

Trinidad and Tobago's Multicultural and Transnational Legacies

Trinidad's society is multi-culturally diverse, with ethnic and racial intermarriage and inter-ethnic/racial births being such a historical legacy that ethnic mixing through the generations is the rule rather than the exception. Trinidad's colonial and plantation legacies, intra-regional influxes and waves of extra-regional immigration over five centuries have all resulted in a heterogeneous mix of Afro-Trinidadians, Indo-Trinidadians or East Indians (from South Asia), Chinese, Venezuelans, French Creole planters (*petit békés* from the French West Indies), British colonial expatriates, Syrian-Lebanese, Portuguese Madeirans and 'small islander' British West Indians from Barbados, Grenada and St Vincent and the Grenadines. In recent times another British colonial and post-colonial source of immigrants has been the South American 'West Indian' territory of Guyana (formerly British Guiana) – with Indo-Guyanese and Afro-Guyanese selectively seeking haven in Trinidad from their troubled, Forbes Burnham-inflicted 'failed state'.

The Republic of Trinidad and Tobago is rich in natural resources, and it has successfully developed its human and natural resources to enjoy one of the highest per capita incomes in Latin American and the Caribbean – with a GDP per capita (PPI) of $19,700 (2006 estimate). Trinidad is the larger and more developed, more urbanized and economically diversified of this twin-island nation. Tobago, for long an ignored dependent partner with a much slower pace of life and a considerably lower overall standard of living, has begun developing its tourism potential, which has recently brought about unprecedented changes. For example, Tobago is now receiving growing numbers of Trinidadians and European expatriates, the latter coming predominantly from Germany, Britain and Scandinavia (St Bernard 2006).

Trinidadian Transnational Networks, Experiences and Practices

Such has been the global spatial diffusion of the Trinidadian diaspora that the development of transnational social networks and transnational family support systems is not only well advanced, but is global and multi-local in character (Conway and Potter 2007; Potter et al. 2008). Many (though we don't know the pervasiveness of the practices) contemporary transnational Trinidadian families' transnational social networks have multi-local sites, or 'transnational spaces', where members can find numerous opportunities for educational advancement, skill acquisition, among other kinds of overseas experience which build human and social capital stocks (Conway 2005; Jackson et al. 2004; Levitt 1998). Family members across generations, frequent North American, European and other dispersed locales, and make use of family ties, connections and 'addresses' as social cushions, opportunity fields, and familial support systems (Chamberlain 1995, 2006). Permanent emigration, more temporary circulations, repetitive to occasional return visits, returns to retire and eventual permanent returns 'for good' are common mobility strategies of those involved in such well-entrenched transnational networks that link Trinidadians with family, kin, partners, friends, 'aunties' and other extended family (often, across generations), in various intra-regional and extra-regional nodes of this diaspora in Barbados, Grenada, Jamaica, Canada, America (the US), the UK and places farther a field, (De Souza 1998, 2005, 2006; Ho 1991, 1993). And drawing upon their own 'narratives' and 'migration stories', it is the interplay between the return experiences and transnational backgrounds of our sample of prolonged sojourners that represents the specific focus of this chapter.

Objectives and Themes

In this section, a short methodological background is provided, describing how such an 'invisible' sample of returning nationals was sought out and interviewed. Following this, addressing this 'return migration of the *next generation*' in the context of Trinidad and Tobago, we present our narrative-based findings on the experiences, practices and views of a selective cohort of 24 relatively young returning nationals or 'prolonged sojourners' as we style them here, We do this under the following headings:

i) Why did they go overseas?
ii) What has motivated this *next generation* of prolonged sojourners to return?
iii) Do they maintain their transnational identities?
iv) How smooth has their re-adjustment been on return?
v) Are they staying, re-returning or moving on?

Qualitative Methodology and Modes of Analysis

Until quite recently, research on Trinidad and Tobago's return migration and the influences transnationalism is having on these circulators' lives has been relatively sparse. Ho (1991, 1993, 1999) and Chamberlain (2006) have focused on transnationalism in particular, while Lee-Cunin (2005) and de Souza (1998, 2005, 2006) have analyzed returnees' experiences. Recently, the authors of this chapter have made contributions to this welcome growth of interest in Trinidad and Tobago's contemporary migration experiences (Conway et al. 2008; Conway and Potter 2007, 2008; Potter et al. 2008; St Bernard 2006), of which this chapter is a continuation.

The research in this chapter utilizes the detailed narratives provided by 24 prolonged sojourner[1] returning transnational migrants to Trinidad and Tobago. The interviews were semi-structured and were carried out between October 2004 and July 2005. Having located a number of such returnees during preliminary field visits, snowball sampling was employed thereafter. The interviews were carried out by the three authors and one trained Trinidadian female assistant. They were generally carried out in peoples' homes, except in the few instances where informants preferred to meet in more public places such as restaurants and bars in central Port-of-Spain. With the authors having carried out interviews of a comparable nature in Barbados, Tobago and St Lucia, we were confident that our identities did not adversely affect the willingness of the informants to talk openly about their early-to-present experiences and their views on migration and return, the adaptations they had made and future intentions for re-migration, or re-return. Most of our informants were in their mid-careers, in the mature-age cohorts of their 30s and 40s. Many were going through the stages of immediate family formation and consolidation, so that we were 'intercepting' them at a mid-point in their life-course trajectories.

Subsequent to the fieldwork, all of the interviews were fully transcribed. The analysis of the informants' narratives was primarily carried out manually – that is – by reading and coding for themes and notable passages. Thus, the extracts presented have been selected as examples of commonly expressed views among our informants – 'common narrations'. In those instances where the narratives selected are indicative of unique or idiosyncratic perspectives, then this is made clear in the text. When introducing the narrators of their extracts, brief mention is made of their background and the contexts in which experiences and practices occurred, where it is felt to be helpful.

1 The 'prolonged sojourner' returnees in our original sample of 40 stayed away for lengths of time ranging from 12 to 30 years. After acquiring more education and work experiences, some got married, had children, built their careers or businesses, and generally settled down 'over there'. Others divorced, or did not marry, but built their professional profile in anticipation of their eventual return 'some day'. Returning in mid-career and mid-life course stage – in their 30s and 40s – their return home obviously completed an international circulation, and in between there may have been visits, which were repetitive 'returns' followed by 'going away again' back to their metropolitan transnational locale.

The Return of prolonged sojourners

(i) *Why Did They Go Overseas?*

Among our informants, almost all left to go overseas to acquire higher education, to seek better opportunities, improve their level of education, or to study abroad. Two circulated during their early 20s, the first trip being to acquire education, the second trip to continue or begin a career 'over there'. The ages they left varied between the late teens and early 20s. Most finished their courses, and continued to stay abroad to acquire experience, pursue further education (completing Masters' programmes or post-graduate medical training), and to gain professional training that would improve their career prospects, whether the latter was to be achieved in the US, Canada or the UK, or in Trinidad. Common views on this were expressed in the following extracts:

> Well, the first consideration is with respect to improving my standard of education and as such trying to improve the situation in terms of my getting a so called 'good job'. As a young man, you have been taught that if you want to be well remunerated you should be well qualified. So I took leave of Trinidad to go to Canada, so as to improve my educational status and at the same time, in an attempt to improve my earning capacity.

> I always wanted to further my studies, so I had the opportunity to go abroad and do that. My parents actually couldn't have afforded that in Trinidad and I was awarded a scholarship, so I went abroad on the scholarship and also to study, which was a great experience for me.

> Primarily, it was to get an education and broaden my experience. I felt as though I was spinning my wheels in Trinidad, and I had graduated my 'A' Levels without any feeling of direction, or what I wanted to do, and I felt that going abroad would give me more options and present more challenges and help me to grow.

Two of our informants undertook two circulations prior to returning to Trinidad in their mid-careers. Denise,[2] a prolonged sojourner, left for North America in the late 1970s to further her education, then returned. Later, tired of her job in Trinidad and looking for something different and new, she then left again for career advancement. Her temporary status during these circulations proved to be a major reason why she returned both times:

> I have a problem living in somebody else's country and not having my papers and I don't like being called a migrant, a foreigner. On both occasions when I

2 The names of informants are pseudonyms.

> left, there was never any thought of, or any consideration of, my not coming back. It was to go for a purpose and to come back home.

Trevor also made two separate circulations. The first time he left to attend medical school in Jamaica (for three years), then returned and completed his course in T & T before leaving again for the UK for another 10 years to get post-graduate medical training there. Family matters intervened in Trevor's itinerary, however:

> The first time I left Trinidad was in the mid-eighties when I went to Medical School in Jamaica, and I stayed there for three years, and I did not intend to come back. But, my wife's grandmother was ill and that is why I came back. But I really enjoyed medical school in Jamaica. I thought it was of a high standard. I would have liked to have completed all of it, there.

Then, after completing his studies, Trevor moved away again, this time to the UK, where he stayed for 10 years and gained more professional experience before again returning to Trinidad:

> I was actually quite annoyed with Trinidad when I left, because I told them we should have a postgraduate training programme, but we did not, and that is why I left.

(ii) What Motivated This Next Generation of Prolonged Sojourners to Return?

The decision to return home was always an option for our informants, because none had fled to avoid incarceration, indebtedness, arrest, or other legal indiscretions that might complicate, or discourage their return. However, most of our informants did not return immediately at the successful conclusion of their university studies, their course completion, or failure to finish and drop-out. None experienced such financial difficulties, or health problems that they returned prematurely, or even as planned, after graduation. Having invested considerable social capital in these 'away-from-home' places, what influenced them to return is now examined next through the lens of our informants' 'own words'.

Five informants – Denise, Michael, Elton, Charles and Sandra – all indicated they had always harboured a desire to return, with Elton citing his parents as influential in this respect, Sandra declaring she was home-sick and Charles indicating that in addition to his long-held desire to return, he felt it was also best for his children and their upbringing. For these prolonged sojourner native sons and daughters, going away was rationalized by an accompanied insistence that a return would eventually be possible, and necessary.

Charles felt very strongly about his social obligations and ties to his homeland, so that return for him was never a question of 'if', but 'when':

> Well, it has always been my desire to work in Trinidad, to live in Trinidad, to bring up a family in Trinidad because all I knew was Trinidad and even with my experience in the metropole, I still felt I had a bonding to Trinidad, the home of my birth. I feel to myself that I am a patriotic citizen; I really love my country and even though I find the life in Canada was more amenable for advancement and opportunity, yet, I still had that yearning to be back once I had qualified. Because I always feel that once I have qualified enough I would be able to give back I would feel better that I am contributing to my society, rather than a developed country – they wouldn't feel any input from me *per se* because they are already developed. I feel I would be able to contribute something and feel proud of it that I can contribute to the development of my country.

Several of our informants insisted that they had returned in response to parental needs; such as a single mother seeking her daughter's return, or the desire for overseas children to be closer now they were older, or problems such as illness that necessitated a return of the migrant or his spouse to help look after elderly or sick parents. Carol, a 'prolonged sojourner', who returned with two children and a 'bi-cultural' (foreign) husband from the UK, had this to say about her reason for returning:

> For the children to live in Trinidad and Tobago … it was for the children, and of course family and brothers and so on helped … My husband was really influential in that respect, because he was interested in getting the children to live in this other culture.

Returning from Canada, Monica had this to say about the 'pull' of family, especially her dad, and the 'push' of the Canadian winters:

> I guess you can say after spending 19 years abroad in a very cold country, and most of my family being back in Trinidad, that was the compelling factor for me to come back. My family was here, and I got a little bit tired of the cold weather, I wanted to be back in the hot weather …
>
> … Indirectly they played a role, because they are part of the reason that I came back – but directly they didn't really play a part in coming back … For the most part, it was my dad who sort of prompted me actually. He asked me if I would consider coming down, and perhaps I was considering coming down just for two weeks or so, or maybe a month, so I guess I would say my dad instigated [the return decision] by asking a question.

Returning from the UK, Julia also commented on her parents' guidance and assistance, although it was a mixed message:

> Kind of half and half … I think in their hearts they wanted to have their child home with them, to spend time with them and that kind of stuff. On the other side, in terms of guiding me, they were not so sure in terms of me coming home, whether it was a good thing for me to be coming home or not in terms of what I want, professional or career wise, that kind of thing.
>
> They said we would love to have me home … I mean it was the kind of thing, if you come home we would love to have you. But if you decide – we will totally understand and think it is probably better for you not to come home. In terms of assistance, I suppose … I am living with my parents now. So they have … [*inaudible* – put up with me?] … in the home. For the past year of me being here in Trinidad I have been living with them. So they have been that kind of support where I don't have to go out and find somewhere else to live.

The eventual return of others was more influenced by immediate family concerns, by husbands, or dependent children's perceived needs. Another reason was the presence of a partner, or spouse-to-be in Trinidad, while they were abroad and trying to maintain a long-distance relationship, or together trying to come to a decision on which one of the pair would move, and deciding that returning to Trinidad would solve this conundrum.

Joseph, a Trini-Canadian, whose father was Indo-Trinidadian and mother was Canadian, had this to say about his 'transnational relationship' and how he and his girlfriend resolved it by his return:

> Well, there were several reasons. The primary reason was my girlfriend of several years was still here, so we had made a decision to get married, and so on. So I came back for that reason, primarily. But I believe I would have come back regardless of that, because I felt that eight years abroad [after four years of university] is enough … It is a cold country in many ways, and I felt that I had a better cultural connection to here. Life was a lot slower paced here, and I felt that I was ready to utilize some of the skills that I had acquired abroad here.

James, on the other hand, had a much briefer 'close encounter' during a return visit to Trinidad that dramatically changed his Canadian life plans and prompted his return:

> Well, after studies I worked in Canada and I came back to Trinidad on a vacation once and I met a young lady who is now my wife and believe you me that is the reason why I came back to Trinidad … I decided I would stay here rather than take her out of Trinidad and go to live in a new country. I thought it would be better if I were to relocate because I already know Trinidad.

For Trevor, on the other hand, it was his wife's 'hatred' of England that definitively, influenced their return, irrespective of his own ambivalence to his medical work and his considerable frustration with working conditions in hospitals in Trinidad. He did, in addition, have ageing parents living in Trinidad, which also influenced his return decision:

> Well, I used to think – here I am doing all these things and it is not my people, that was one, and the other factor was my wife hated the country, she did not like England at all, so to preserve my marriage I had to come back. If I had stayed she would have come back. And I would have had to have a 'come-between' marriage, which I do not believe in …
>
> … My parents are important to me because my parents were now alone and getting old, so I thought that it was important that I come back.

(iii) Do They Maintain their Transnational Identities?

There are several overlapping questions to ask, and answer, about our informants' transnational identities. Did they 'keep in touch' by repetitive visiting while away? And, do they continue their transnational practices – such as repetitive visits and circulations, and use of IT to remain in touch with friends and communities left behind? Do they maintain their transnational identities – duel-citizenship, keeping in touch with overseas friends, continuing their transnational visiting – now they have returned? Again, a selection of our informants' narratives addresses each of these questions about the influence of transnationalism.

Repetitive visiting – 'keeping in touch' prior to return Typical of the ease with which many of our informants' moved back and forth after going away to university, Laura, a returning prolonged sojourner from the USA described her repetitive visiting pattern this way:

> In the early stages, missing home a lot and so on, I used to come home practically every summer, and every Christmas vacation. After about two or three years, I became fairly more established, I got a job while going to school, so the visits home became less frequent. I remember there was a period for about three years I had not returned because I was working during the vacation, and then after that about once every two years or so I would make the trip.

Some, of whom Michael was quite typical, moved back and forth regularly, either once a year or twice, with Christmas and Carnival being favourite times to reunite with family, and enjoy themselves 'playing mas'. Michael recalled most of his earlier regular visits:

> I used to come to Trinidad either once a year or twice a year. So it's a considerable amount of time or visits that I made to Trinidad. I made it

> a duty to be here every year and then I stopped coming. There was a long lapse. After '83, I didn't come back until '87. Then I came back in '90, '91, then I came back in '02, '03.

When asked about his record of visiting, Garth, a prolonged sojourner returnee coming back from the USA after 24 years away, had this to say:

> Every year. With the exception of 1980, I went out there. I came back in 1982 Summer, and then when I was doing my Green Card and all that stuff, I probably came home in 1987, and from 1999 onwards, every year, maybe a couple years I did not show up but for Carnival every year. Last year because of buying the house and stuff, I was here like four or five times last year, which gave me a better perspective on Trinidad outside of Carnival, and maybe see how the country ticks outside of Trinidad.

For others, financial constraints hindered their repetitive visiting, so that they stayed away for lengthy periods, and this was the case for Charles, who went to Canada:

> Well, while I was studying I never returned to Trinidad. I hadn't the money to do that, the luxury of finance for travelling backward and forward, I have never had it. I wanted every cent to use.

Some did not make any return visits, because they had decided to give Trinidad a clean break and were not interested in 'keeping in touch'. They had all the intentions of staying away and of permanently emigrating to their chosen overseas metropolitan society. However, as things turned out, some changed their minds, or family pressure changed it for them, so that they eventually returned. Gene, a returning prolonged sojourner from the US described his 'intention not to return' then his 'footlooseness', and finally his change of plans, this way:

> I didn't want to come back so I didn't come back until the very last year after I finished [college]. I never came back. Once I left I didn't come back. It is after I finished by degree and then I came back for a little while and then I went to Jamaica. I came back for about a couple weeks then I went to Jamaica. I didn't want … I had no intentions of coming back to Trinidad. You see when I come back here all the old influence start back again. Your mother telling you listen, do it and come back to Trinidad and work in Trinidad. And my brother say, listen, nobody to see about your mom in Trinidad, you know, so you better go back down there, because you have nobody else again down there to look after her. So, all of that influenced me to come back here.

One of our informants, Susanne, was a prolonged sojourner who had left for London, to 'do some travelling as well as do some studying abroad'. Susanne admitted she came to love living in that city because of its lifestyle and cosmopolitanism. She bluntly told us she did not come back from the UK for a visit, because she didn't want to. Yet, she did return, because of family obligations and affections:

> At the time it was not so much that I wanted to come back, as it was that things were going on with my family in terms of – like bad things – like my grandmother was very ill and then my brother had been in a really bad accident and I was … It seemed like I was falling out of the loop of things. I think there is a certain amount of time that you can be away from your home before you are out of the cycle of how things run. You know, things were happening and I didn't know about it – like little things, not big things, everyday things that you used to be a part of and I think a part of me just wanted to come back home and be with my family for a while and make a decision about what I wanted to do. So it wasn't so much that I wanted to come back as it was that circumstances in my family sort of pushed me to return.

Furthermore, she found adjusting to Trinidad, when she did finally return difficult:

> The first year of my being back was the worst year of my life, literally. No exaggeration, it was the worst year of my life. You get accustomed to the freedom that you always knew existed and then you come back and you realize that … nothing has changed. And not so much in terms of physical change. Yes, physically, things have changed: there are bigger buildings and there is a new government. There are more fast food outlets. But in terms of where change matters: in the people, in the society, in growth, there was no real growth.

Keeping in touch with family and friends 'here' and 'over there' Repetitive visiting certainly helped a few of our informants to keep in touch both with their parents and Trinidad, their birthplace. It was also instrumental in familiarizing these absent nationals to 't'ings back home', so that their adjustment and adaptation experiences on return were commonly eased and facilitated. Thus, returning from England, Julia certainly found her visits had helped:

> I think I was reasonably well prepared because I visited; you know, I was coming back all the time. So it is not like I was coming to a totally alien place that I didn't know anything about; didn't know how anything worked, and I didn't have any support mechanisms. I didn't have to go and make totally

> new … all new friends. I felt I had a few people; my parents and I had one or two friends. So I think, on a scale of one to 10, I would say I would probably have been a six or a seven in terms of knowing what I was coming up against … There have been lots of things.

On the other hand, a few of our informants either did not visit, or declared that visiting was not at all helpful, nor did it greatly influence their decision to return, or prepare them for the cultural and societal shocks which accompanied return. For Carol, visiting was disappointing and often a turn off, because of the difficulties she experienced during exploratory visits with her young children and foreign-born husband. On the other hand, James, who returned from Canada, did not find his repetitive visiting had any influence on his decision to 'come back':

> Well, Trinidad was my home and I visited Trinidad every summer I had vacation, while I was abroad studying … It really didn't influence my decision, tho'. The experience was not influenced by my visiting; it was influenced by the mere fact that I wanted to come back home and be in Trinidad with my girlfriend at the time.

Ida, a prolonged sojourner returnee from the USA, had also visited Trinidad regularly, but the experience had neither particularly influenced her final decision to move back home, nor had it influenced her adaptation: 'not particularly' was her dismissive assessment. On the other hand, Laura a prolonged sojourner returnee from the USA, who stayed away for 17 years, but returned as a 'single mom' with her baby to be closer to her mother, had this more reasoned explanation to offer on the subject:

> I think the visiting experience was always pleasant, and I think that creates a sense of fantasy. I know now that it certainly creates a sense of fantasy with regards to what life is really like in Trinidad. It was always pleasant, it was always liming; everybody is on vacation so everyone is spending time with you, the partying, and what not. So the vacation is always pleasant. Even if unpleasant circumstances are going to happen let's say with my mother or something, I think it is always in both of our minds, that I am only here for just a couple of weeks, I am going back soon and I would not have to put up with this … or she is going back soon, so I would not have to put up with this. So, it really does create a huge fantasy, for life in Trinidad isn't easy. It certainly is not the American way. It really creates a fantasy as to what it is really like to live in this country … I am thinking about planning to go back. I am thinking seriously about going back.

Keeping in touch via email and phone Almost without exception, all but two of our returning informants maintained close ties with the circle of friends they left behind in North America or the UK; their overseas transnational community. Family – siblings, favourite aunties or closest, estranged parents – living elsewhere, were included in some returnee's circles, but among such communications networks, email proved to be the most common means for regular 'keeping in touch', with the phone being considered an expensive alternative for a few. Regular visiting by some of our informants, who had maintained business connections, also helped reinforce the transnational connectivity that was valued by almost all of our prolonged sojourner returnees. Only a minority, however, had kept their property, or house, or foreign bank accounts open, and a few had deliberately got rid of everything as a final gesture before 'returning for good'. One informant regretted doing this, but others insisted they did not need to maintain such connections or hold on to such investments.

Dual citizenship: a strategic advantage Sixteen of our informants maintained dual citizenship rights and dual identities in that they possessed two passports, one from Trinidad and Tobago, the other from the country in which they had spent considerable time acquiring higher education and professional skills – Canada, the US and the UK. Three of the female returnees, insisted they had 'dual identities', feeling they belonged in the US as much as they belonged to Trinidad and Tobago. However, none of the three have formal citizenship rights, although one holds a Resident Alien 'green card'. Rather they identify with the US through their husband or their children, and appear to be quite content with this arrangement that facilitates travel back and forth and allows them to retain transnational contacts with friends and family by visiting in person. One of these, Kathy, is a prolonged sojourner returnee from the UK, who does not have dual citizenship, but insists she has a transnational identity, being half-British/half-Trinidadian:

> When you say dual identity that is different from dual citizenship? You see one could have a dual identity, you can be half British in your thinking … Well, I don't have any dual citizenship. It might be dual identify … What I mean is – I don't just think Trinidadian, I think much more widely. It is something that happens to you after you are away for a long time.

Joseph is a prolonged sojourner, who spent a considerable time abroad in Canada. His father is an Indo-Trinidadian, his mother is white, Canadian by nationality. They are divorced, and his father is now back in Trinidad and Tobago, while his mother is in British Columbia. Concerning his dual identity, Joseph admitted he was born with the advantage of dual citizenship:

> My mother being born in Canada and so on, when I was born, immediately before I was a couple years old I had two passports. So I grew up like that, moving with that flexibility. I believe it is important to maintain that, because

> in the event, or in the unlikely event of anything catastrophic happening, you always have somewhere else to go.

Carol a prolonged sojourner who returned with two children and a bi-cultural (her description) foreign husband, acquired dual citizenship through her marriage. Carol said she maintained her dual identity, kept her ties with her previous overseas country – the UK – but also kept her Trinidadian connections and ties as well:

> I always did it when I was abroad. When I was abroad I kept the Trini ties, and since I have been here I am keeping the other ties.

Dual citizenship, therefore, for the majority of our informants has proven to be an advantageous transnational identity. Breaking the mould, somewhat, a couple of our informants, Elton and James, were quite comfortable with giving up their dual citizenship on their return, declaring that one passport, and one national identity was enough for them, now that they had 'returned for good' (for more on this topic, see Conway et al. 2008).

(iv) How Smooth Has their Re-adjustment Been on Return?

Again, several overlapping questions qualify in terms of adjustments and experiences on return. Have these mid-career returnees encountered problems of adjustment, since they have been away for more than a decade or more? Or has return been 'no problem'? Does their adjustments and reception by those who stayed behind depend upon the social and family institutional networks they access, and how does their economic 'success' influence their adjustment and acceptance? Do they experience different processes of adjustment in the workplace and social spheres? How has their transnational experiences helped or hindered their adaption? Three general themes emerge, which serve as an organizational framework for letting our informants tell us how well, or how poorly, they adjusted on return.

No problems with adjustment on return The record is mixed, and our informants were nearly split on whether they encountered few, or no problems, or whether the first six months or year was an unexpected trial. Among, the latter, some translated this lack of adjustment into a wish, or intention to re-return, or re-migrate in the near future. Some, however, admitted to initial problems, but then settled down and had no intentions to move again, or at least were relatively content and satisfied with their return decision.

Typical of returnees who insisted they had no problems are Ida, Garth and Julia. Ida had no problems adjusting on her return from the USA, because she prepared herself for this:

> When I went to America, I learned things about new cultures, but I never really strayed from my culture and … I still think that my culture and the morals and values that I grew up with are the things that made me decide where I would choose to stay, whether here or there.

When asked whether her reception depended upon the social and family networks she could access, Ida re-affirmed that her parents and friends neither helped or hinder her:

> No, not particularly. I was prepared for my return while I was out there [in the USA]. I was working hard and knowing that I don't think I would want to live there for ever; did the things I was supposed to do and prepared for my transition.

And Garth had this to say about his preparedness for resettling in Trinidad and Tobago:

> I think I am prepared, because I did not make this decision in two months, it is almost a five year decision, that just sort of gathered speed within the past two years, when I really put things together. So I think from a perspective of understanding Trinidad as compared to living in the US, I understand it is a big difference, and I think mentally I prepared myself. The physical part is not hard; the mental part is just accepting that certain things are done differently here. I was kind of prepared for that, for instance like getting a job in IT, the pay scales are going to be different, stuff like that. So my expectations I guess were realistic, coming back to Trinidad soon. I think I was really prepared to settle, and so far it has been comfortable.

Julia contributed this on how well she adjusted to life back in Trinidad:

> I think I was reasonably well prepared because I visited; you know, I was coming back all the time. So it is not like I was coming to a totally alien place that I didn't know anything about; didn't know how anything worked, and I didn't have any support mechanisms. I didn't have to go and make totally new … all new friends. I felt I had a few people; my parents and I had one or two friends. So I think, on a scale of one to 10, I would say I would probably have been a six or a seven in terms of knowing what I was coming up against, or knowing what I would have to deal with, and knowing also the joy that I would find, because there are some things that have been wonderful to come home to.
>
> But the quality of life is much easier in Trinidad. Yes, things are much, much slower, and I still find myself working very, very, hard. But the quality of life is better. You don't have to rush around from here to there to everywhere in the freezing cold all wrapped up on the tube or struggle

> with lugging groceries back or anything like that. Those kinds of things, you don't have to do that. So the quality is better. Getting a car and going to the grocery, and you know, make your groceries and then come home … that sort of thing.

Problems in the economic sphere, few in the social sphere Sandra, a prolonged sojourner returning from the USA, expressed confidence in her overall transnational preparedness, but found her adjustment at work more difficult than her social reception:

> Trinidad was no problem; probably because I was back every year, it never made a difference. So there weren't any problems adjusting. And similarly in the States I never had a problem.

She did, however, experience different processes in the workplace and social spheres:

> Yes, in the workplace – only because it was a completely different environment, which we go through the first couple weeks of just getting used to the place and getting to know people. But as such it wasn't like culture-shock or anything. I think probably it was because of the career I was in.

Successful adjustments had been made by most of our informants on return, but a few either could not adjust, or found Trinidad had changed for the worse. Typical of this minority opinion, Vera had problems with the increases in crime and security issues in Trinidad:

> I thought I was very well prepared. I really thought I had all my bases covered and as I got here I realized, no. The driving for one I found quite scary, you know, when I arrived and so on. And seeing things like personal safety you know, making sure you lock the doors and the car doors. When I left Trinidad we used to go up to Lady Young [Highway Overlook] and look out, you know. When it was Halley's Comet, we used to wake up at four in the morning, you know, with your binoculars and go up to Lady Young. Now, you can't do that … So, you know, just the change with the crime and so on. Social changes, I was not prepared. Practical things, yes, I was.

Family networks and support most effective When deciding which social networks had been most influential in both their decision to return and how well they adapted on return, factors such as family support, family influence and the presence of parents, siblings and immediate family turned out to be the most common factor among our informants. Typical of the comments on this, Charles put it this way:

> As I say, my family really played a major part in my life and whatever happened to me in terms of my academic career, whatever happened to me in terms of what I have developed over time, in terms of training and in terms of assessing things in a different way, whatever way I may want to look at it, I think that my family they have always been there for me.
>
> A large role was played by my mother. She always insisted that we must be educated to the max. She felt that it was necessary; that you never settle until you feel you reach the pinnacle of your career. And knowing that she had raised 10 children and knowing the situation of poverty and that thing, she always felt that what she hadn't acquired during her time, she always wanted to see her children do well and perform at the maximum whatever their path. So I think the encouragement was very much given from my mother, in particular, my brothers and sisters too – my other brothers also encouraged me to leave to study.
>
> … So in leaving and in fulfilling, it has always been the joy to return back to those who encouraged you and you want to be a part of them because you want them to see how well their son has done and how much their son knows and all that kind of thing, so that they will feel proud of their son. So that is the kind of thing with it. So I got a lot of encouragement from my parents.

Carol had this to add, though much more briefly and to the point:

> Well once we came back people were encouraging and helpful … my friends were as helpful and encouraging as they could.

(v) Are They Staying, Re-Returning or Moving On?

After spending many years abroad, all of our pre-retirement, informants have returned to the island of their birth in mid-career and mid-life course stage in their 30s and 40s. The question we now turn to concerns whether they are staying for good, thereby completing their migration cycle, only staying for a while or are uncertain about their future migration plans, or are likely to re-migrate and re-return to the metropolitan society they evacuated, because they found social adaptation too difficult.

It turns out that there is no consensus among our 24 informants, as the narratives that follow clearly demonstrate.

Uncertainty reigns On her return with her immediate family, Carol found adjustment to Trinidad a problem, but she was still contemplating her options, and still retaining her flexibility to consider re-migrating, when we interviewed her:

> My adjustment … has been so complicated on return, getting focused back into something, it took some time. So, the fact that we had a little to keep us

> afloat, that helped because I think if we were not able to do that, in fact, we would not have even bothered to come if we did not have a little something to play with. So that helped because it took eight months to be professionally established … working.
>
> … There are enough reasons to be wavering about returning or not, but put it this way, it is too early. As a reasonable human being you would give it time. It is too early to make a decision one way or the other.

Carol felt she needed to keep her re-migration options open, for several reasons:

> You see the context is not as clear cut when your family has a mixed background – one partner is from the continent and then the children are bi-cultural, so we will really leave the options open in that context.
>
> … In any case, even in terms of two Trinis who have lived abroad for the same number of years, they still need to think, because most of them, their kids were born abroad.

Gene, also, was adamant about keeping his remigration options open and flexible, but at the same time felt he needed to stay on account of his children's upbringing – 'chained down' is his description:

> … So, right now, if I had the opportunity, I would go where I got used to …
>
> I would go back up. If I didn't have all the commitment with my kids I would leave. I would leave right away. I don't mind we living in a small place out there. I know I could do something up there. I know I could make something up there. But I have other commitments, and because of that, so that's why I am chained down here … I could migrate tomorrow. Plus, you see, with all the crime, you now getting fearful of even going out and party. You know, getting fearful even to go to a function outside; to go to a dinner outside. Because you fear that you go to some place and somebody stick you up in the car park. It is getting like that.

Julia had this to say about her social adaptation and whether she had plans to re-migrate:

> It depends on which day you ask me. I change … You know, it just depends on the day. I don't think … social adaptation generally has been so bad that it is forcing me to leave Trinidad, no.
>
> Yes, but … in the job that I aspire to, that I want to do, that I need to get more exposure that I can't get in Trinidad. And therefore, long term planning would not be to remain here for the next five years … I need to be experiencing things in different places in order for me to progress.

> I mean, I plan to be here for a while – I don't know how long is a while. … But yes, like I said, I think I don't want to be here forever. I think I also don't want to leave and never come back. I don't want to leave in two years and that be it and not come home to Trinidad to live … I would want to come back in the future. But I just don't want to be here from now until then. I think I have a lot more learning to do, and because of the job that I do, I need to be outside … I don't have to go back to London. I am very open to living anywhere.

And, influenced by how she felt that her life-path journey was still to be charted, Monica had this to say about adapting to life in Trinidad on her return, and whether she might stay or re-return to Canada (she has dual citizenship). Reflexively, her single state appeared to play a role in her ambivalence towards staying:

> I cannot see myself settling here solely. I used to think that Trinidad was a better place to raise a family, and even though I am single still, I am not so sure that Trinidad really is a better place. In regard to the extended family concept where you have the support of your 'Tantie' or Uncle or who you have or Granny, more so than putting your child in a day-care abroad, that support is here, that network is here. But besides that, I really don't see a whole lot keeping me here. So it is just a matter of time.

Staying for good Elton eschewed dual citizenship and stressed coming back for good and 'doing good':

> Nah, nah, I ain't going nowhere. I had a green card and all that kind of thing and I choose to give it up … I didn't want the hassle of them questioning you: 'why you out so long'? I just chose to give it up.
>
> I am coming back home. I am Trinidadian. As I said earlier, I always try to keep in touch. Every opportunity I had to come home I would come. So it was just a – the time away from home was just for developmental reasons. In a way you have to, you know, thinking about coming back home to help Trinidad and Tobago, and this present administration with the Vision 2020, bring this country to developed country status. So, I mean, if I could be home and be playing a part of that … That's how I always feel, coming back to help Trinidad and Tobago.

Susanne, on the other hand, was much more ambivalent about her situation and her migration intentions. However, in the final analysis she was pragmatic in defence of her choice to return from the UK, while retaining the need to keep her future migration options flexible:

> Well, every day I think about leaving, not because I am not happy but just because I can get a lot more done faster. But you see, you have to choose. And that's why I talked about – and that basically is my belief about the power of choice, about choice being really important; the power of the individual to create their own happiness. I choose not to, because away, because of the circle I was in, I would invariably add to wealth. You know, the circle I was in, most of the people I knew were wealthy people, a lot of really big business men, and because of where my restaurant was, in the West End, we were around a lot of high-up market places. So invariably we end up mixing and meeting with people who have resources. And invariably if I decide to get out of one industry and go to another, I would have ended up in very much in the situation where I would have been adding to wealth and not creating it. And here, one has to create wealth.

Susanne goes on to express her strong opinions about how the brain drain is depleting Trinidad's creative stocks of human capital – the country's youth (though she doesn't use such terms):

> And here, in Trinidad, the problem is that we have no one creating wealth. Older people are in government, are in corporations that are headed by directors and CEOs who have been alive for longer than we can think of. There is no real creation of wealth and there is no real creation of avenues for young people. So, many of the young people who have talent and who have the ability to do things, in order to move forward, are leaving. And it sounds very clichéd, but I've always wanted to do something for my country …
>
> … So I choose to create – or try to at least – set up a company and create avenues and/or create wealth for people who don't have it and thereby helping my generation of people … So, I choose to try and be an instrument of change, even if it is in a small way. Even if I help fifty people, at least fifty people would have found a way that wasn't there before. And I think that's all you can do, really.

Susanne concludes with some thoughts on her role as a professional return migrant, while at the same time including transnational practices in the flexible options she envisages for her future:

> In Trinidad, I am in the process of setting up my own company in such a way that I would have the opportunity to leave and travel. I can live here as long as I have the chance to run away three or four times a year for a while or twice a year for a month or maybe three four times a year for like a week, or two weeks. As long as I can do that I'll be fine, because I need to get refreshed there and I need to get new ideas …Right now, I am in the process of setting up my own company, *Future Opportunities Abroad.*

> My old boss is still asking me when am I coming back, you know. So I know that is an option. But, also, there are other options as well because there are more opportunities opened in other European countries … And, as globalization continues to take place and we become like literally one world, I suppose it would be more.

Conclusions

By returning in their 30s and 40s, our returnee informants have demonstrated they are prolonged sojourners rather than permanent emigrants, so their connections to their birthplace and homeland might be expected to be influential with regard to their 'development potential' and their willingness to return home to help 'make a difference', and 'give something back'. Running through our informants' narratives is the importance of family ties and family influences on their decision-making, and their return reasoning. Most left to improve their human capital, acquire higher education and professional training and experience, to take advantage of opportunities that they did not feel existed in Trinidad and Tobago. Many used transnational connections, or the transnational backgrounds of their family, to facilitate their movement away, and to help them gain access to university courses and to employment and study visas in Canada, the USA and the UK. Differences in each of these destination's immigration policies also influenced our informant's routes to these overseas opportunities.

Parental guidance, support, and occasional firm directives were instrumental in many of our informant's transnational lives in general and in their migration decision-making in particular. Mothers were singled out by several informants as being the stronger influence on persuading informants to return, but for a few, their fathers were financially important in financing return visits, helping them become established back home, and giving advice. Concern for their children's upbringing, in which Trinidad was considered in highly favourable terms, was also a commonly cited reason that several of our informants gave for returning and staying. Among other informants with career hopes and plans, Trinidad was not without its problems, in part because of work-place difficulties, and in part because of the deteriorating social climate brought on with the alarming and unsettling rapid rise in violent crime and robbery in the country.

Just about all of the prolonged sojourners we interviewed had very positive things to report about their transnational lives and experiences while away from home, but most were also careful to keep in touch with things back home. Many visited regularly while abroad – once or twice a year – or wished they could if it had not been due to financial limitations. Repetitive visiting and regular short trips back home ensured they kept in touch with family and friends, and also enabled them to keep pace with changes that were occurring while they were absent. For some, this pre-return, visiting strategy helped them adjust and adapt, when they returned, and encouraged them to return. For others who regularly visited, it did not have much influence either on their return decision, nor their social adaptation. Work-place difficulties were faced by many of our informants, in part because of the informant's comparative lens working in two different work climates and cultures; one away in North America or the UK, the other at home in Trinidad and Tobago. Such difficulties frustrated some of our informants to the point that they were ready to leave, but others accepted the initial culture shock, felt they had moved on and were content enough to stay, even if it was only 'for a while' and not 'for good'.

Acknowledgements

The generosity of the National Geographic Society's Committee for Research and Exploration in funding this research is gratefully acknowledged. We also wish to thank the Trinidadian and Tobagonian informants who gave freely of their time in talking to us about their transnational lives and livelihoods, their migration and adaptation experiences abroad and at home, and in being so willing to share their thoughts and views on the widest possible range of 'Trini' matters and concerns.

References

Chamberlain, M. (1995), 'Family narratives and migration dynamics: Barbadians in Britain', *Nieuwe West Indische Gids*, 69: 253–75.

Chamberlain, M. (2006), *Family Love in the Diaspora: Migration and the Anglo-Caribbean Experience*, New Brunswick, NJ: Transaction Publishers.

Conway, D (2005), 'Transnationalism and return: "home" as an enduring fixture and an anchor', in Potter, R.B., Conway, D. and Phillips, J. (eds), *The Experience of Return Migration: Caribbean Perspectives*, Aldershot and Burlington, VT: Ashgate, 263–82.

Conway, D. and Potter, R.B. (2007), 'Caribbean transnational return migrants as agents of change', *Blackwell Geography Compass*, 1(1): 25–45.

Conway, D., Potter, R.B. and St Bernard, G. (2008), 'Dual citizenship or dual identity?: does "transnationalism" supplant "nationalism" among returning Trinidadians?', *Global Networks*, 8(4): 373–97.

De Souza, R.M. (1998), 'The spell of the Cascadura: West Indian return migration', in Klak T (ed), *Globalisation and Neoliberalism: the Caribbean Context*, London: Rowman and Littlefield, 227–53.

De Souza, R.M. (2005), 'No place like home: returnees R & R (Retention and Rejection), in the Caribbean homeland', in Potter, R.B., Conway, D. and Phillips, J. (eds), *The Experience of Return Migration: Caribbean Perspectives*, Aldershot and Burlington, VT: Ashgate, 135–56.

De Souza, R.M. (2006), 'Trini to the bone: return, reintegration and resolution among Trinidadian migrants', in Plaza, D. and Henry, F. (eds), *Returning to the Source: The Final Stage of the Caribbean Migration Circuit*, Jamaica, Barbados, Trinidad and Tobago: University of the West Indies Press, 74–104.

Ho, C.G.T. (1991), *Salt-Water Trinnies: Afro-Trinidadian Immigrant Networks and Non-Assimilation in Los Angeles*, New York: AMS Press.

Ho, C.G.T. (1993), 'The internationalization of kinship and the feminization of Caribbean migration: The case of Afro-Trinidadian immigrants in Los Angeles', *Human Organization*, 52(1): 32–40.

Ho, C.G.T. (1999), 'Caribbean transnationalism as a gendered process', *Latin American Perspectives*, 26(5): 34–55.

Jackson, P., Crang, P. and Dwyer, C. (2004), 'The spaces of transnationality', in Jackson, P., Crang, P. and Dwyer, C. (eds), *Transnational Spaces*, London and New York: Routledge, 1–23.

Lee-Cunin, M. (2005), 'My motherland or my mother's land?: return migration and the experience of young British-Trinidadians', in Potter, R.B., Conway, D. and Phillips, J. (eds), *The Experience of Return Migration: Caribbean Perspectives*, Aldershot and Burlington, VT: Ashgate, 109–34.

Levitt, P. (1998), 'Social remittances: migration driven local-level forms of cultural diffusion', *International Migration Review*, 32(4): 926–48.

Potter, R.B., Barker, D., Conway, D. and Klak, T. (2004), *The Contemporary Caribbean*, London and New York: Pearson/Prentice Hall

Potter, R.B. and Conway, D. (2008), 'The development potential of Caribbean young return migrants: "making a difference back home ..."', in van Naerssen, T., Spaan, E. and Zoomers, A. (eds), *Global Migration and Development*, London and New York: Routledge, 213–30.

Potter, R.B., Conway, D. and St Bernard, G. (2008), 'Transnationalism personified: young returning Trinidadians, "in their own words"', *Tidschrift voor Economische en Sociale Geografie*, 100(1): 101–13.

St Bernard, G. (2005), 'Return migration to Trinidad and Tobago: motives, consequences and the prospects of re-migration', in Potter, R.B., Conway, D. and Phillips, J. (eds), *The Experience of Return Migration: Caribbean Perspectives*, Aldershot and Burlington, VT: Ashgate, 157–82.

St Bernard, G (2006), 'Episodes of return migration in Tobago: a phenomenological study', in Plaza, D. and Henry, F. (eds), *Returning to the Source: The Final Stage of the Caribbean Migration Circuit*, Jamaica, Barbados, Trinidad and Tobago: University of the West Indies Press, 188–213.

Chapter 10
Returning Youthful Nationals to Australia: Brain Gain or Brain Circulation?

Graeme Hugo

Australia is one of the world's great immigration nations, with 24 per cent of its 21 million population (2006 estimate) born in a foreign country, and with a further 26 per cent who are Australia-born but have at least one parent who was born overseas. Yet it is less well known that Australia is a significant country of emigration with a diaspora of around 1 million expatriates (Hugo 2006a). Moreover, because Australia is one of the very few nations which collects comprehensive data on people who leave the country as well as for those who move to the nation, it is possible to obtain a comprehensive picture of both emigrants and immigrants. These data show a high and increasing loss of young, skilled Australians which very much reflects Australia's peripheral position in the global economy as well as the strengthening of globalization processes and internationalization of labour markets. Indeed the exodus has raised some concerns in Australia that the loss of skilled human capital is causing a brain drain (Wood 2004). This has been countered by those who point out that immigration levels in all major skill areas are well above emigration levels (Birrell et al. 2001). However it is less well known that this brain drain is being offset by considerable return migration among the Australian expatriate population and it is this return migration which is examined in this chapter.

The chapter begins with a discussion of the sources which are utilized. These include the emigration data collected by the Australian Department of Immigration and Citizenship (DIAC) referred to earlier, as well as a number of surveys and qualitative in-depth studies of expatriate Australians. The major patterns and trends of emigration from Australia are then outlined and the characteristics of emigrants examined. The extent of return migration among the Australian diaspora is then considered, followed by an analysis of the motivations prompting return. An assessment is then made of the impacts of return migration and, finally, some policy implications are discussed.

Data Considerations

It is unfortunately true that countries around the world generally collect comprehensive data on immigrants and immigration, yet very few collect equivalent information on those leaving the country (Dumont and Lemaitre 2005; Schachter 2006). Accordingly, immigration is studied to a much greater extent than emigration and, as Ley and Kobayashi (2005, 112) point out: 'The weight of the assimilation narrative, especially in the United States, has tended to obscure the significance of the return trip home.' Australia, however, is an exception. The same information is collected on the flows of people who leave the country as for those arriving. The following questions are asked of all people leaving the country: country of birth, date of birth, gender, occupation and destination. Then for three categories of departures the following questions are posed:

a. *Visitor or Temporary Entrant Departing*: (i) length of residence in Australia; (ii) state of residence in Australia; (iii) country of residence.
b. *Resident Departing Temporarily*: (i) intended length of stay abroad; (ii) main reason; (iii) country in which will spend time abroad; (iv) Australian state of residence.
c. *Resident Departing Permanently*: (i) ccountry of future residence; (ii) Australian state of residence; (iii) if not born in Australia – how long ago did you come to Australia? – did you intend to settle permanently?

On the basis of (a), (b) and (c) persons leaving the country, foreign visitors, residents and citizens alike are categorized into three groups: visitors, long-term and permanent departures. Those involving Australian residents and citizens can be defined as follows:

- *short-term movers* – Australian residents and citizens whose intended stay abroad is less than 12 months;
- *long-term out-movements* – departures of Australian residents and citizens who intend to return; with the intended or actual length of stay abroad being 12 months or more;
- *permanent departures* – Australian residents and citizens (including former settlers) departing with the stated intention of residing abroad permanently.

A problem here is that this threefold differentiation is based on people's intentions and these do not always come to fruition. Nevertheless the data do provide insights into the emigration process and in this study we focus on the last two groups. While there is good *flow* data on emigrants it is much more problematical in terms of movement of the *stocks* of emigrants. There have been some attempts to bring together census data from OECD countries to provide estimates of the stock of emigrants for particular countries of origin (Dumont and Lemaitre 2005) but these provide quite partial estimates.

The research in this chapter also draws upon and uses the findings of three studies of Australian emigrants who are resident outside the country. The first of these is a 2002 survey of 2,072 Australian expatriates made up of: (i) 1,327 alumni of Australian universities whose current address was in a foreign country; and (ii) 745 persons who responded to advertisements placed in newsletters of Australian expatriate organizations. The second was an on-line survey of 9,529 Australian expatriates undertaken by a consortium of Australian expatriate organizations to coincide with the Australian 2006 population census (The 'One Million More' Study). A third survey of 1,581 Australian expatriates in the United States was undertaken as part of a PhD thesis (Parker forthcoming). In all three cases, the samples are not representative and are selectively sampled by drawing upon cohorts of overseas Australians who have had some contact with expatriate newsletters and newspapers. It is also selective and biased in that it accordingly accesses the more highly skilled groups. In addition to the three surveys, a number of in-depth interviews with Australian expatriates in the United States, United Kingdom and several Asian countries were undertaken by the author, and their 'narratives' are drawn upon where illustrative and representative.

Emigration from Australia

The United Nations (2006) lists Australia as having the eleventh largest immigrant population of any country in the world and ranking ninth in terms of the largest percentage of its residents being foreign-born. Moreover, as Figure 10.1 shows, the settler inflow has increased in recent years. It is also shown that the numbers of residents of Australia who have left the country on what they reported as a 'permanent' basis has also increased. Moreover, the ratio of emigrants to immigrants has increased from 0.15 in 1988–1989 to 0.52 in 2006–2007. There are two classes of those undertaking permanent departures from Australia – former immigrants who are returning to their homeland or to a third country and the Australia-born (Figure 10.2). The former category of 'settler loss' (and of re-return) has perhaps involved one in four immigrants to Australia during the post-World War II period (Hugo 1994; Hugo et al. 2003). Of more interest here is the Australia-born component of this emigrant flow and their numbers have increased rapidly from 8,399 in 1989–1990 to 20,234 in 1999–2000 and 36,882 in 2006–2007.

However, 'permanent departures' are only part of the exodus of Australians. The outflow also includes a group of Australians who indicate that they are leaving for more than a year, but who intend to eventually return. Figure 10.3 shows that the numbers of Australian residents who have departed temporarily on a long-term basis has increased sharply in recent years. Indeed, the numbers

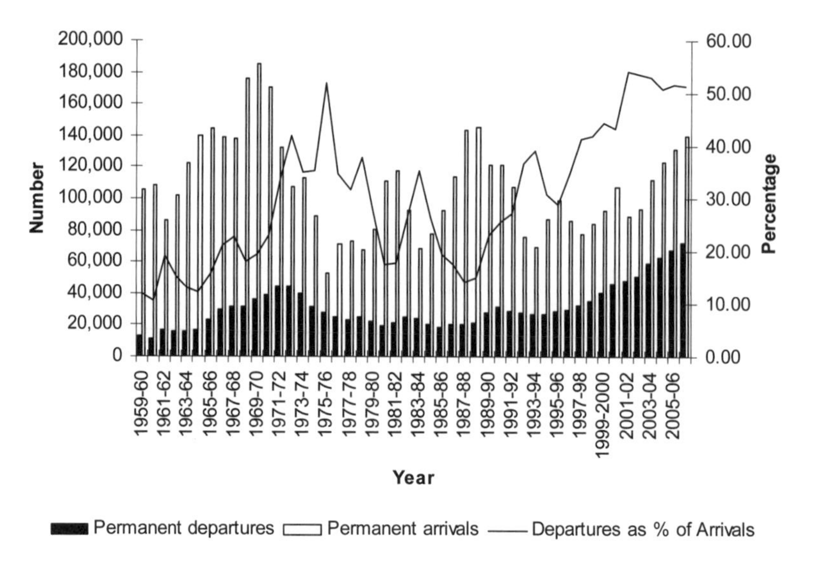

Figure 10.1 Australia: permanent arrivals and departures, 1959–1960 to 2006–2007

Source: DIMIA *Australian Immigration: Consolidated Statistics* and *DIAC Immigration Update*, various issues.

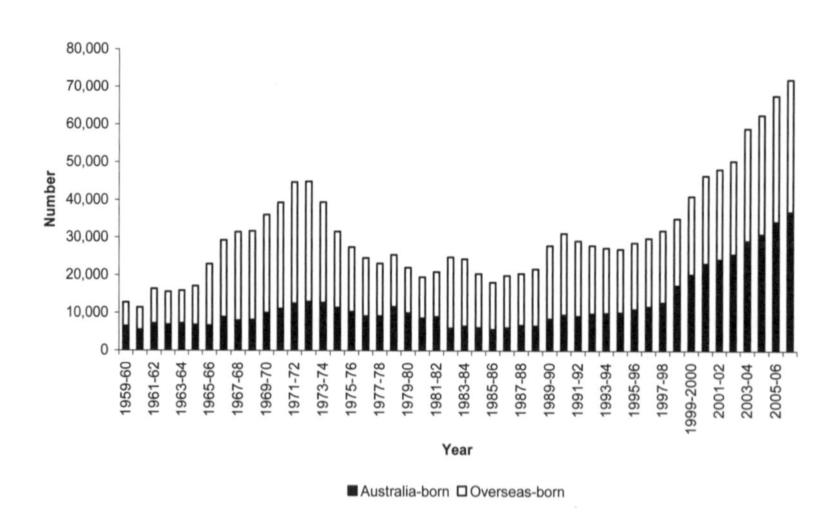

Figure 10.2 Australia: permanent departures of Australian-born and overseas-born persons, 1959–1960 to 2006–2007

Source: DIMIA *Australian Immigration: Consolidated Statistics* and DIAC, *Immigration Update*, various issues.

more than doubled between 1986–1987 (48,854) and 2006–2007 (101,610) and they increased by 38 per cent over the last decade. Between 2004–2005 and 2006–2007 the numbers increased by 10.9 per cent indicating that the rapid growth is continuing. In terms of intended length of stay abroad, the majority (82.7 per cent) intend to stay away for less than three years, but 5.2 per cent indicate that they will be away for five years or longer (Table 10.1).

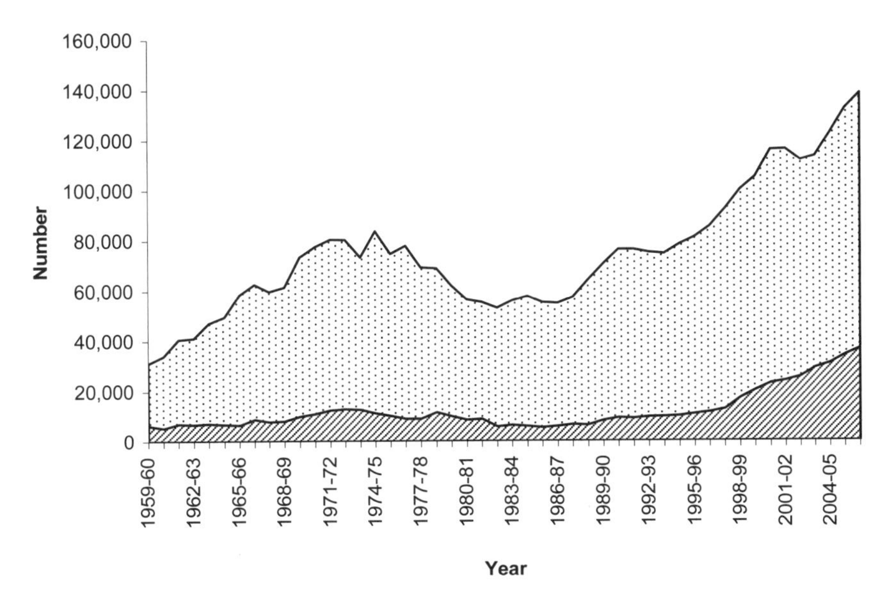

Figure 10.3 Australia-born permanent departures and long-term departures of Australian residents, 1959–1960 to 2006–2007

Source: DIMIA *Australian Immigration: Consolidated Statistics* and DIAC, *Immigration Update*, various issues.

What are the demographic statistics of these Australians who are leaving on a permanent or long-term basis? Like all migrations this outflow is selective by age. Figure 10.4 shows the age structure of permanent departures from Australia over the 1993–2007 period and the youthfulness of the Australia-born component of the outflow is clear; with the modal age group being 30–34 and more than half aged between 25 and 44 (52.7 per cent). The presence of a large number of children aged less than 10 years (18.9 per cent) points to the significance of young families among these Australians who are leaving the country permanently. There is almost a perfect balance between males and females (sex ratio of 99.9) in the permanent outflow of Australians, but

Table 10.1 Australia: long-term departures of Australian residents by intended length of stay, 2004–2005 to 2006–2007

(Intended) Length of stay	Number	Per cent
1 and under 2 years	123,699	42.5
2 and under 3 years	117,880	40.5
3 and under 4 years	29,075	10.0
4 and under 5 years	6,981	2.4
5 or more than 5 years	13,723	4.7
Grand total	291,358	100.0

Source: DIAC, unpublished data.

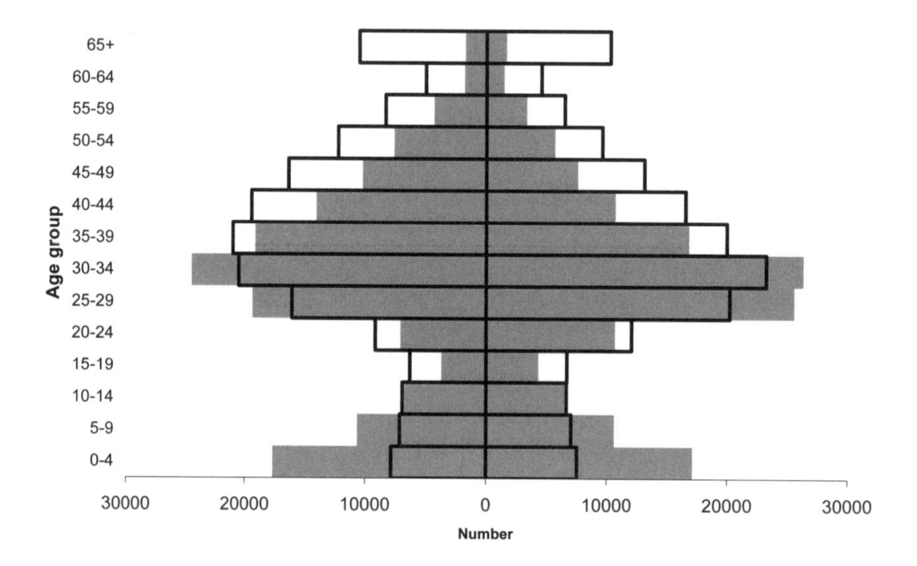

Figure 10.4 Australia: permanent departures, Australian-born and overseas-born by age and sex, 1993–1994 to 2006–2007

Source: DIAC, unpublished data.

females substantially outnumber males in the 20–34 age group (sex ratio of 81.3) with males being more numerous in the older groups.

Turning to the long-term departures, Figure 10.5 shows an even younger profile with 25–29 being the modal age group and 36.5 per cent being in their 20s compared with 21.1 per cent of permanent departures of the Australia-born. Moreover, it will be noticed that there are only small numbers of dependent age children with only 9.1 per cent aged less than 10 years. This indicates that those Australians leaving on a long-term basis are not only younger than their permanent counterparts, but also more are moving as singles or couples rather than as families with dependent children. There are more males than females (sex ratio 102.7), although females outnumber males in the 15–29 age group (sex ratio 83.9).

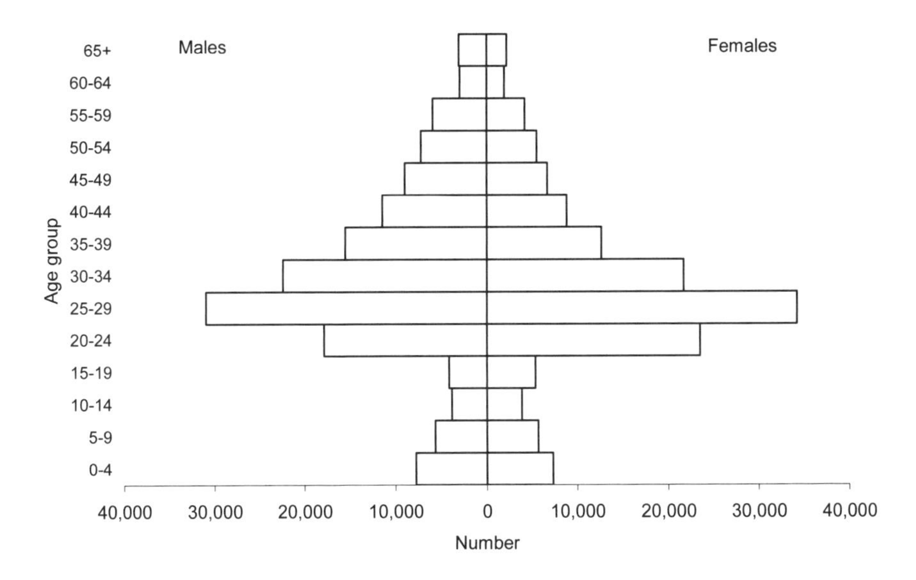

Figure 10.5 Australia: long-term departures – Australian residents by age and sex, 2004–2005 to 2006–2007

Source: DIAC, unpublished data.

In addition to their youth, long-term and permanent emigrants from Australia are selectively drawn from the more educated and higher income groups. Table 10.2, for example, compares their occupational structure to that of the total population of Australia at the time of the 2006 population census. Strong concentrations occur in the managerial and professional occupations. It is apparent that the labour markets for these types of jobs are now international. Furthermore, in such labour markets many of the most sought-after jobs are in major global cities and Australians' participation in them necessitates international migration. Moreover the favoured destinations (the UK and USA account for 36.1 per cent of permanent departures of the Australia-born and 40.1 per cent of Australian resident long-term departures) are selective regarding to whom they will grant a work permit or temporary or permanent residence and because entry requirements are usually based on employer-supported skill and education needs.

Table 10.2 Australia: permanent and long-term departures by occupation and total population by occupation, 2006

Occupation	Departures		Total Australia 2006 Census
	Permanent	Long-term	
Managers and administrators	19.2	13.8	9.2
Professionals	45.5	48.8	19.6
Associate professionals	9.8	9.3	12.2
Tradespersons	5.1	6.7	12.3
Advanced clerical, sales, service	4.1	3.3	3.2
Intermediate clerical sales and service	13.0	14.6	17.2
Intermediate production and transport	1.0	1.2	8.2
Elementary clerical sales and service	1.5	1.3	9.6
Labourers	0.7	1.0	8.5
Total	100.0	100.0	100.0
	24,630	51,834	8,938,579

Source: DIAC, unpublished data and Australian Census of Population 2006.

Return Migration

While it is clear from the above that there has been a high and increasing level of emigration of skilled Australians in recent times, there is also a high level of return migration among those leaving on a permanent or temporary, long-term basis. The extent of return migration among permanent departing Australians is not immediately apparent because Australian citizens make up only a small fraction of those designated as permanent arrivals among those entering the country. Between 1993 and 2007 there were 1.39 million settler arrivals and of these only 20,409 (1.5 per cent) were Australian citizens. Moreover, when their age structure is examined, (see Figure 10.6) these arrivals are strongly concentrated in the dependent child age groups – more than two thirds being aged less than 5 years and 85.7 per cent aged less than 15 years old. Clearly, these are predominantly the children of Australians born while their parents were overseas who are coming to Australia for the first time when their parents return.

However, it appears that most of their parents are classified as long-term Australian resident arrivals rather than permanent settler arrivals even though when they departed many designated themselves as undertaking a permanent departure. This is apparent when we examine the long-term arrivals of Australian residents. Figure 10.7 shows that the arrivals of Australian residents after 'long-term' absences of more than one year closely track

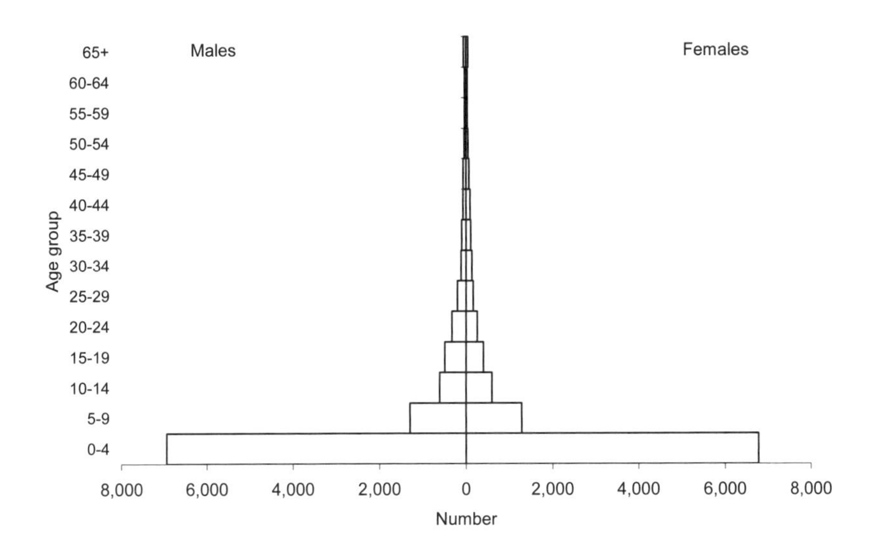

Figure 10.6 Australia: settler arrivals, total and Australian Citizens by age and sex, 1993–1994 to 2006–2007

Source: DIAC, unpublished data.

Figure 10.7 Australia: Australian resident long-term arrivals and departures, 1959–1960 to 2006–2007

Source: DIMIA *Australian Immigration: Consolidated Statistics* and *DIAC Immigration Update*, various issues.

the level of long-term departures over the last five decades, with numbers increasing substantially in recent years. Moreover, in recent years arrivals have outnumbered departures. This is a strong indicator that the number of Australian resident long-term arrivals includes not only people who were designated as Australian resident long-term departures when they left Australia, but also many who were considered *permanent* departures when they left. Further evidence of this pattern is provided in an internal study undertaken within Australia's Department of Immigration and Citizenship (DIAC). Since 1999, all persons entering or leaving Australia have been allocated a Personal Identifier (PID); making it possible for analysts to follow the subsequent travel movements of these individuals. Using this facility, Osborne (2004) identified all persons who had ticked the 'Australian Resident Leaving Permanently' box on the departure card between July 1998 and June 2003 and matched them against all subsequent movements recorded by their PID. He found that 24 per cent of them in actuality had returned to Australia permanently, although they previously had indicated that they were leaving permanently, when departing from Australia. Only 59 per cent of original permanent departures could be designated definitely as having left permanently. Hence, there is a significant disconnect between many emigrants' stated intentions and actual behaviours. Presumably, though not measurable, there is also likely to be a similar disconnect among people who intended leaving on a temporary long-term basis, but have, in actuality, remained overseas permanently. Nevertheless, and regardless of the latter under-estimation of those permanently staying overseas, there is clear evidence of substantial return migration not only by the Australian resident long-term departures but also the Australia-born permanent departures.

The age-sex composition of Australian resident long-term arrivals is shown in Figure 10.8 and the most striking feature of it is, like the departures, its youthful age structure. It has very small numbers in the dependent child age groups, but as was shown above, it seems likely that many of the children of returning expatriates who were born overseas were classified as 'settlers' rather than returning residents after long-term absences. In Table 10.3 the age structures of long-term arrivals and departures of Australian residents are compared and the younger age structure of the departures is in evidence with 59 per cent aged under 30, compared with 46 per cent of arrivals. Moreover, while there was an overall net loss of only 2,084 between 2004 and 2007, as would be expected with an essentially circular movement, among young Australians aged less than 30 there was a net loss of 28,141, which was more than offset by a net gain of 30,225 in older age categories. This appears to indicate that long-term residents undertaking temporary moves abroad are tending to move out in their 20s and staying on average two or three years so that the returnee population is older than the long-term departures.

However, although these returning nationals are slightly older than the Australian departures it is important to stress that they are still overwhelmingly

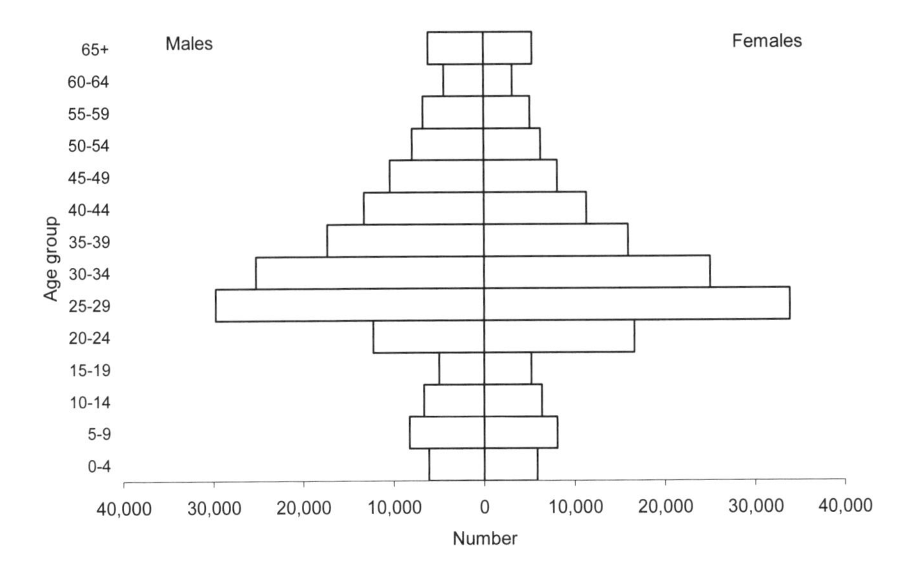

Figure 10.8 Australia: arrivals of Australian residents who have been away for more than a year, by age and sex, 2004–2005 to 2006–2007

Source: DIAC, unpublished data.

in the younger, pre-retirement working age groups. Reviewing the Caribbean literature on emigration, circulation and return, Conway and Potter (2007: 35) have posited:

> To this point, extant scholarship has commonly focused on elderly, first generation retirees who have consummated their long-held desires and intentions to return to the homelands of their youth' and who have left metropolitan Britain, Europe,, or North America for good.

This characterization, however, does not appear to be appropriate in the case of contemporary Australian returnees. Only 6 per cent are aged 60 years or over, and among these, are an important group of older former migrants who circulate between their original homeland and Australia, as is indicated by a relatively significant proportion of long-term departures of the Australia-born being in this age group (3.5 per cent, 10,126 persons). The returnee Australian population is overwhelmingly then a young adult onc with 55.8 per cent being aged between 20 and 39 years old. There is little difference in the gender balance between the permanent resident long-term arrivals (102.3) and departures (102.7) with males being slightly more numerous than females in both.

The countries from where these nationals have returned to Australia are depicted in Figure 10.9 and a widespread pattern is in evidence. The

Table 10.3 Australia: long-term arrivals and departures of Australian residents by age, 2004–2007

Age group	Arrivals		Departures		Net migration
	Number	Per cent	Number	Per cent	
0–29	143,743	46.0	171,884	59.0	-28,141
30–30	83,494	26.5	72,320	24.8	+11,174
40–49	42,993	13.6	35,990	12.4	+7,003
50–59	25,970	8.2	22,836	7.8	+3,134
60+	19,040	6.0	10,126	3.5	+8,914
Total	315,240	100.0	313,156	100.0	+2,084

Source: DIAC, unpublished data.

United Kingdom and, to a lesser extent, the USA are the major origins. Of particular interest, however, are the large numbers returning from Asian countries. This reflects the dynamic economies of the Asia-Pacific region, which have attracted large numbers of skilled expatriates to jobs where there are insufficient highly-trained locals to fill them. Also noteworthy, the strong circularity in trans-Tasman migration between Australia and New Zealand (Bedford, Ho and Hugo, 2003) is reflected in the large number of returning nationals (Australians) from New Zealand.

The main countries of origin where the Australian residents were living before returning after a long-term absence are shown in Table 10.4. Not surprisingly, the predominant 'destinations-turned-sources' are countries which play important roles in the global economy; many with major global cities (and financial centres). Hence 41.3 per cent were previously in European Community countries, with three quarters of them in the United Kingdom. Another 12 per cent returned from North America – the United States and Canada – while there were substantial numbers who had been living and working in the global Asian cities of Hong Kong and Singapore (10.9 per cent) as well as the rapidly growing economy of China (4.7 per cent). Of course, some of this return movement involves former immigrants who became Australian citizens and left later to spend an extended period in their homeland before re-returning to their adopted country (Australia), but the bulk of the circular movement of nationals seems to flow between Australia and neighbouring New Zealand and between Australia and a dispersed network of major global economies.

As would be expected, the returning Australians are a highly skilled group. Indeed, it might be confidently assumed that they are generally returning with enhanced skills and experience as a result of their overseas experience. Table 10.5 shows that half of all returnees were in the professional occupation category compared with a fifth of all resident workers. Almost three quarters

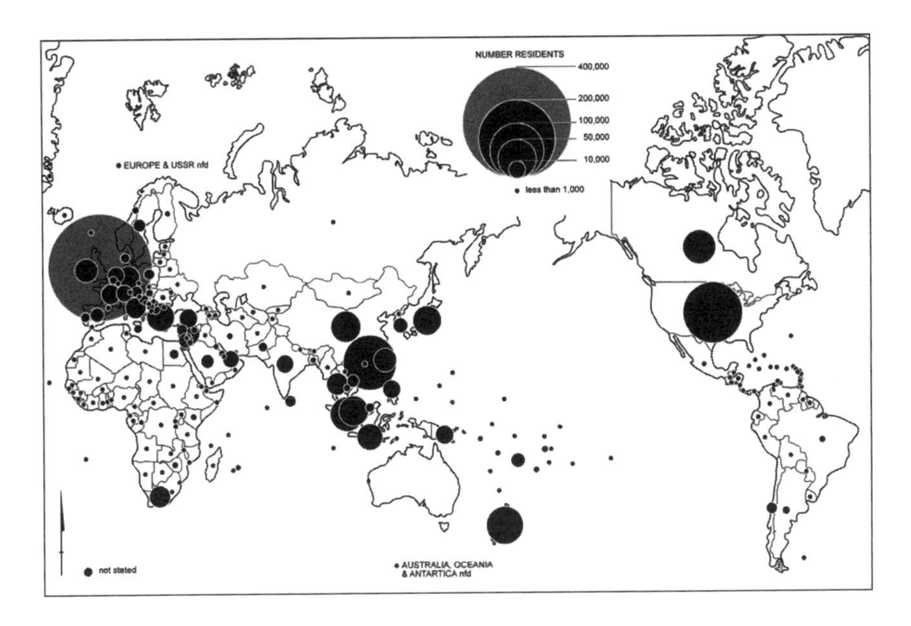

Figure 10.9 Australia: country of previous residence of returning Australian long-term migrants, 1993–1994 to 2006–2007

Source: constructed from unpublished DIAC data.

Table 10.4 Australia: permanent resident returnees after a long-term absence – country of previous residence, 2006–2007

Country	Number	Per cent
United Kingdom	34,301	31.2
United States	9,835	8.9
Hong Kong	7,496	6.8
China	5,235	4.7
Singapore	4,547	4.1
New Zealand	4,203	3.8
Canada	3,370	3.1
Malaysia	2,755	2.5
Japan	2,484	2.3
Other European Country	11,148	10.1
Other	29,067	27.1
Total	110,041	100.0

Source: DIAC, unpublished data.

(72.8 per cent) are in the three highest status occupations in the Australian system, which compares more than favourably with the 33.6 per cent of the total Australian workforce estimated by the 2006 population census. Hence, these returnees are a significant (though neglected) part of the nation's annual immigration intake which is contributing to the growth of human capital in Australia.

Table 10.5 Australia: arrivals of long-term Australian residents by occupation, 2004–2005 to 2006–2007 and total population by occupation, 2006 (per cent)

Occupation – major group	Long-term residents returning		Total population 2006	
	Number	Per cent	Number	Per cent
Managers/administrators	24,244	12.1	818,105	9.2
Professionals	99,130	49.4	1,748,219	19.6
Associate professionals	22,671	11.3	1,090,715	12.2
Tradespersons	15,221	7.6	1,102,537	12.3
Advanced clerical and service workers	5,992	3.0	288,837	3.2
Intermediate clerical, sales and service workers	24,687	12.3	1,536,818	17.2
Intermediate production and transport workers	2,714	1.4	736,331	8.2
Elementary clerical, sales and service workers	4,388	2.2	858,671	9.6
Labourers	1,816	0.9	758,346	8.5
Total	200,863	100.0	8,938,579	100.0

Source: DIAC, unpublished data and Australian Census of Population 2006.

These data clearly show that there is a strong pattern occurring of young Australian skilled professional workers leaving the country and spending several years working in major global cities expanding their skills, experience and professional networks before then returning home. Moreover survey work among Australians overseas indicates that a firmly held intention to return to Australia is common. Table 10.6 shows that in three surveys of Australian expatriates a high proportion indicated that they intended to return. The highest rate was recorded in the large 'One Million More Survey' which was global in its coverage and the least in the United States-based survey. It is interesting to note that in the 2002 Australian Emigration Study which was global in its coverage, respondents in the United States and Canada were the least likely to intend to return (44.8 per cent) of all those interviewed. Alternatively, there is a relatively high proportion who were undecided about whether they should/

Table 10.6 Surveys of Australian expatriates: intentions to return to Australia

	Australian Emigration Study 2002	One Million More Survey 2006	US Expatriates Study 2007
Intend to teturn	50.7	63.5	35.7
Do not intend to return	17.2	10.9	30.8
Undecided	32.1	25.6	33.5
Total Per cent	100.0	100.0	100.0
No.	2,072	8,744	1,581

Source: Hugo et al. 2003, 50; One Million More Survey; Parker forthcoming.

would return to Australia or not. This reflects a considerable degree of conflict which many expatriates articulated during in-depth discussions. On the one hand they wanted to return to Australia to be closer to family or for lifestyle reasons, while on the other hand the professional and job opportunities open to them in the destination country were pulling them in the other direction. This tension is evident in some of the statements made by the interviewees:

> Many of us overseas are desperate to return home, but there are very limited opportunities and salaries are less than half what they are here in the US.

> I'd just like to say that I wrote undecided [to the intentions to return question] even though I know I'll move back to Australia. But at present we don't know when or how as it depends on employment plus in order for my husband to get a visa I will have to go there before my husband does and perhaps live and work there for more than 6 months.

Nevertheless there were also many who strongly expressed a desire to return:

> I love Australia and want to return and hopefully make a contribution although I will need to compromise my career to do so.

> We live in Silicon Valley, California as do numerous other Australians (thousands). Of the hundreds that I've met the majority intend to return to Australia, most commonly within a 10 year span.

One of the defining features of Australia's diaspora, which emerged in the three survey studies, was the tension which many expatriates felt between a desire to further their careers, maximize income, etc. overseas and the undoubted conviction concerning the family and lifestyle advantages of their homeland. This often means that many expatriates attempt to maximize the advantages of both locations by spending extended periods in foreign countries during their 20s and 30s and early 40s and then returning to Australia as they enter the early family formation stages of their life cycle. To many of these expatriates it is important that their children spend at least part of the formative years of childhood in Australia.

Reinforcing this point, analysis of the Australian arrival and departure data shows that there is a high level of return migration among Australians who live and work in foreign countries. Moreover, it is clear that a majority of the returning expatriates are in their early middle working years and are highly qualified, experienced and well-connected workers who have considerable potential to act as 'agents of change' in Australia and make a significant contribution to the Australian economy. Conway and Potter (2007) argue that most of the extant literature on return migration focuses on retirees who are generally not considered effective agents of change, in part because of their age, but also due to their exit from the formal labour market and the adaptation problems they face back home, due to their lengthy sojourns overseas.

Again, this characterization does not apply to most Australian returnees and those still overseas who have intentions to return to Australia. The Australian arrivals/departure data analysed earlier point to the fact that the bulk of Australian return migration is of people in the 25–39 age bracket; indicating that many return not so much at the end, but rather at the peak, of the career cycle. And, the importance of 'return during mid-career' is backed up by the results of the three surveys of expatriates (Conway et al. (Chapter 9) also find this to be a common pattern among Trinidadian 'transnational sojourner' returnees). Data drawn from the three Australian expatriate surveys in Table 10.7 show that high proportions of those still in their early career-years intended to return to Australia within five years of their interviews.

Motivations for Return

It is clear that return migration among Australian expatriates is substantial (and increasing) in scale, and that it does not at all conform to the 'now-open-to-question' stereotype of returnees being predominantly of people nearing the end of their careers, or returning on retirement. The Australian expatriate population is overwhelmingly highly educated, skilled and young. They have travelled overseas in their 20s or early 30s in search of career advancement in an increasingly globalized world in which head office activities of large, increasingly powerful, multinational companies are more and more concentrated in far-flung, global economies (natural resource extractive industries, for example) and global cities. Allied to this is a strong desire to experience living in a different cultural context, a sense of adventure and a wish to extend life experience. They travel overseas as singles, couples or as a couple with very young dependent children. Among the reasons for leaving

Table 10.7 Surveys of Australian expatriates: when respondents intend to return to Australia by age

Australian Emigration Study 2002 (n=1,044), per cent						
Plan to return to Australia	**Age**					
	Less than 35	**35–49**	**50+**	**Total**		
Within 1 year	17.9	13.7	13.8	15.8		
1–2 years	25.2	17.4	16.7	21.2		
Longer	57.9	68.9	69.5	63.0		
One Million More Survey 2006 (n=5,554), per cent						
Plan to Return	**Age**					
	20–34	**25–44**	**45–55**	**55+**	**Total**	
Within 1 year	18.6	11.0	10.1	11.4	15.8	
1–5 years	64.0	54.0	47.0	54.6	59.8	
5–10 years	12.1	22.0	27.5	22.5	16.3	
More than 10 years	1.0	4.6	8.4	1.1	2.5	
Don't know	4.3	8.4	7.0	10.0	5.7	
United States Expatriates Survey 2007 (n=565), per cent						
Plan to return	**Age**					
	Under 30	**30–39**	**40–49**	**50–59**	**60+**	**Total**
Less than 1 year	28.0	13.3	12.0	7.0	23.5	15.3
1-2 years	14.0	18.4	16.0	21.1	5.9	16.9
2-3 years	14.0	16.4	7.2	10.5	11.8	13.2
3-5 years	18.0	11.7	8.8	15.8	23.5	13.0
More than 5 years	10.0	17.2	22.4	29.8	5.9	19.1
Undecided	10.0	23.0	33.6	15.8	29.4	22.5

Source: Hugo et al. 2003, 50; One Million More Survey; Parker forthcoming.

which were given by respondents to the three expatriate surveys (see Table 10.8), there is a strong emphasis on higher income, job opportunities on the one hand, and on the other hand a keenness to experience new different and exciting things in new cultural contexts.

Interestingly and noteworthy, however, these reasons for "seeing and experiencing the world beyond Australia" contrast sharply with the reasons given by expatriate respondents for intending to return to Australia. Table 10.9 shows that there is a clear emphasis on family and lifestyle in their responses, with only a small minority indicating that work-related factors would attract them back to Australia. On the other hand, the reasons given by those who intend to remain in the destination (Table 10.10) place a high degree of emphasis on employment opportunities, salary, career progression etc.; although, it is also true that those who have partnered with a national at the destination and whose children had grown up there were also more likely to be not interested in returning.

Table 10.8 Surveys of Australian expatriates: reasons given for leaving Australia (per cent)

Reason	Australian Emigration Study 2002 n=2,070	One Million More Survey 2006 n=9,529	US Expatriates Study 2007 n=1,581
Better employment opportunities	42.6	53.9	81.4
International experience/experience new culture	36.1	62.7	43.3
Job transfer, career advancement	23.7	19.4	44.2
Family, marriage	22.3	17.4	36.8
Partner's employment	12.1	11.1	13.0

Source: Hugo et al. 2003, 50; One Million More Survey; Parker forthcoming.

Table 10.9 Surveys of Australian expatriates: reasons given by respondents for intending to return to Australia

		Australian Emigration Study 2002	One Million More Survey 2006	US Expatriates Study 2007
Lifestyle		82.9	65.3	61.2
Family		71.5	77.5	73.5
Work		15.7	14.9	12.7
Education		9.8	5.4	4.8
Total	Per cent	100.0	100.0	100.0
	No.	1,050	5,637	1,529

Source: Hugo et al. 2003, 50; One Million More Survey; Parker forthcoming.

Table 10.10 Surveys of Australian expatriates: reasons given by respondents for remaining at the destination

		Australian Emigration Study 2002	One Million More Survey 2006	US Expatriates Study 2007
Better employment opportunities		45.4	37.8	48.5
Career opportunities		40.8	42.0	49.0
Higher income		40.4	44.2	44.7
Marriage/family		38.6	19.3	46.0
Lifestyle		30.6	–	17.7
Total	Per cent	100.0	100.0	100.0
	No.	1,022	974	997

Source: Hugo et al. 2003, 50; One Million More Survey; Parker forthcoming.

Among Australians there is a strong life-course dimension to return migration just as there is to the original emigration. Many seek to return to Australia in their late 30s and 40s. This is arguably when they are at the peaks of their careers with a decade or more of experience behind them in their profession, during which time they have developed extensive professional networks and expertise. Given this situation it has to be asked why such privileged expatriates would seek to return when their human capital value is high, their earning capacity is nearing its highest and they have built up the foundation for enhancing their status as part of a transnational elite. The answer appears to be linked very strongly to social capital formation and its significance to these mobile Australians. It has already been shown that the reasons given for return by expatriates overwhelmingly relate to family and lifestyle factors and are not strongly related to economic and career factors. This of course does not mean that the latter are unimportant. Indeed, gaining a job in Australia that is satisfying and well-remunerated is a fundamental, *necessary* condition of the return. However, this job 'back home' may not be as well paid, as transnational or as 'central' as the one held in the diaspora, that will be given up or exchanged. It is social factors which provide the *sufficient* condition for return. A crucial element in the timing appears to be either entry into the early stages of family formation, or passage to a life-course stage where children have reached the end of elementary school age. Respondents firmly articulated this by emphasizing a mix of dependent-related, factors influencing social capital accumulation within their immediate and extended families: (i) the desire to have their children grow up interacting with the wider extended family of grandparents, cousins, uncles and aunts; (ii) the desire for children to have an Australian education; and (iii) the desire for children to grow up as Australians with an Australian lifestyle.

Children figure in much of the responses and the desire to return is much less among those who remain single, without a partner or without children. Where the partner is not from Australia the strength of motivation for return to Australia is also not as strong as where both partners are Australian. As some respondents have expressed:

> … now that we have become settled, we love all that the City of London has to offer economically and culturally it is hard to leave. London is a city for young people and I know that one day I'm going to want my children to grow up in the Australian sunshine and relaxed lifestyle. I will always be an Australian.

> Since being here I've married a British citizen and had a child. We now believe we will move to Australia as a better place to raise a family; despite the low salaries, high taxes and limited work opportunities on offer.

> We plan to return to Australia … we have a son who was born here in Texas last year and wish for him to have an Australian home and education.

The propensity for young Australians to live and work for an extended period in a foreign country has greatly increased in the last two decades. It is clear, however, that the propensity for such emigrants to return to Australia while still relatively young has also increased over time. Why is it that the propensity to return is so strong among young Australian expatriates? One of the strongest findings from the surveys, and especially from the in-depth interviews, has been the strong identification of the diaspora with their Australian background.

All three surveys asked respondents whether 'they still called Australia home' and the resultant data in Table 10.11 shows that the great majority of respondents still regard Australia as their home. Moreover, it is notable that in each survey virtually all of the young respondents aged in their 20s considered Australia 'home'. Only as ages increased was there a steady decline in those identifying with Australia in this way. This was also strongly associated with the length of time that people had been absent from Australia. There was a regular decline in identification with Australia during the period of time they had been away. Countering this last assessment of identity shifts, however, one of the striking findings of the surveys and the in-depth interviews was the strength of identification with Australia among many expatriates; even among those who had no intentions of returning to Australia to live and who hadn't lived in Australia for several decades. Typical of such an enduring sense of national pride and belonging is this firm conviction by an Australian

Table 10.11 Surveys of Australian expatriates: proportion that still call Australia home (per cent)

	Australian Emigration Study 2002	One Million More Survey 2006	US Expatriates Study 2007
Yes	79.3	62.8	80.4
No	16.7	3.9	6.8
Undecided	4.0	4.2	12.8
I have more than one home	–	29.1	–

Source: Hugo et al. 2003, 50; One Million More Survey; Parker forthcoming.

expatriate in the States, though she admits to transnational ties which now involve two 'homes', too:

> I have my husband and family now here in the USA but all the rest of my immediate family are in Australia – it will always be 'home' but I also have a home here. I will never give up my Australian citizenship.

One of the differences between the modern diaspora and those emigrant patterns of earlier times is in the frequency and intimacy of contact that can be maintained with their homeland while overseas. This, as widely acknowledged, is because of the global revolution in information and communication on the one hand and the cheapening and speeding up of international travel on the other (see Rogers 2004). Australian expatriates receive information about local Australian events, politics and sporting results at the same time as their Australia-based relatives, so their knowledge of the *minutae* of Australian life is comprehensive and current. The new technologies not only allow expatriates to keep up with events in Australia but they facilitate intimate contact with family, friends and colleagues. As Azure (2003, 33) writes in a collection of short essays by Australian expatriates:

> When I open emails from loved ones, I hear the words read to me in their voices.
>
> My heart aches because it is pulled and stretched across seas, across lands, to encompass births, deaths, marriages, first homes, losing a job, gaining a job, major successes, major setbacks. When the phone receiver is replaced I smile in a distant land.

Australians overseas are able to interact with relatives and friends in an intimate way, more than expatriates of earlier generations, and this undoubtedly is a factor in strengthening relationships with home. The cheapening of international phone calls and the development of the internet now means that expatriates can speak directly on a daily basis with friends, relatives and colleagues, whereas in the past such regular intimate contact was simply not possible. It was apparent from discussions with expatriates that electronic communication with 'home' had an important reinforcing effect on their connection with Australia and on their identity as Australians. This regular contact is maintained not only with family but also with former colleagues and friends. Professional linkages are also strong. As a result, they are aware of opportunities in Australia and developments within their fields of endeavour. The ways in which Australians kept in touch with what was happening in Australia among the over 9,000 respondents to the One Million More Survey are shown in Table 10.12. Notably the use of on-line media is striking and reflects the extent to which expatriates are fully familiar with events and developments in Australia. It is especially interesting, however, and with reference to Table 10.12, to note that 44.1 per cent keep in touch through having regular contact with Australian colleagues. This facilitates their return to work in Australia when they are ready to return. It is also interesting and noteworthy, that the proportion maintaining regular contact with colleagues was significantly higher (47.8 per cent) among those with firm intentions to return to Australia than among those intending to stay overseas (33 per cent); while those who were undecided about their future plans to either return or stay were in between (39.5 per cent).

Table 10.12 Ways in which respondents keep in touch with what is happening in Australia while away, 2006

Ways of keeping in touch	Per cent
Regular interaction with family and friends	92.0
Regular reading of on-line media	90.7
Regular interaction with Australian colleagues	44.1
International news channels	35.0
Expatriate organizations	24.1
Australian journals/magazines	20.2
Mailing lists	11.4
Other	6.3

*Source:*One Million More Survey (n=9,529).

It is not only through regular electronic communication with Australia that expatriates are able to maintain their knowledge of Australia and reinforce family and professional linkages. Regular visiting has been facilitated by reduction in the time and financial costs of international travel. This also is evident in the surveys of expatriates. In the 2002 survey of expatriates it was found that most expatriates visited Australia on a regular basis despite the substantial distances involved – with over three quarters of respondents being in North America or Europe. Only a small proportion of respondents (13.6 per cent) had not visited Australia since moving, and these 'non-return visitors' were overwhelmingly recent departures from Australia. And, as an additional confirmation of the aforementioned association, there was a strong positive relationship between the length of time expatriates had resided overseas and the number of visits they had made. Furthermore, the fact that a third of the expatriates surveyed had visited Australia ten or more times is indicative of the intensity of connections maintained with the homeland. Similarly, data from the One Million More Survey shown in Table 10.13 indicates that over two thirds of these interviewed expatriates visit Australia once or more than once every two years.

Table 10.13 Australian expatriates: frequency of visiting Australia, 2006

Frequency	Per cent
More than once per year	12.3
Once per year	27.5
Once every 1–2 years	30.8
Once every 2–3 years	14.6
Once every 3–5 years	8.1
Less than once each 5 years	6.8
Total	100.0

Source:One Million More Survey (n=8,879).

Impact of Return Migration on Australia

Assessing the impact of return migration on Australia is very difficult, since there is no way to identify returnees in standard data collections and there have been no national or special surveys of this group, to date. It is important to note, however, that Australian expatriates have a strong sense of what they can do for Australia while they are overseas. In the One Million More Survey, for example, the respondents were asked whether or not they agreed with a number of statements relating to this issue and Table 10.14 presents the results according to whether or not respondents had intentions of returning to Australia. Their responses indicate that a majority felt that they were ambassadors for Australia while overseas while an eighth indicated they actively assist Australian companies by helping to link them with opportunities in foreign countries. This indicates that the expatriate community can and do play a positive economic role for Australia while overseas (Hugo 2006b).

Table 10.14 Australian expatriates: responses to statements relating to contribution to Australia, 2006

Statement	Per cent agreeing			Total
	Intending to return	Remain overseas	Undecided	
1. I feel I am an ambassador for my country and I promote Australia at every opportunity	65.0	41.7	48.8	58.3
2. When people ask me questions about Australia I am happy to share my knowledge	92.7	88.9	90.8	91.8
3. I am gaining skills and experience that I can take back to Australia with me if/when I move to Australia to live	79.9	20.9	55.7	67.3
4. I am making contacts overseas for other Australians/Australian companies	13.9	8.3	11.6	12.7

*Source:*One Million More Survey (n=8,879).

From the perspective of return migration, however, it is responses to the third question in Table 14 which are especially germane. Two thirds of respondents felt that the skills and experience that they were gaining in their foreign-based employment will be of value when they return to Australia. Moreover, this sentiment is strongest among those who have definite plans to return to Australia. In fact, four out of every five in this group believe that they will bring back to Australia greater skill and more experience than they had when they departed. The awareness of these expatriates that they have enhanced their human capital in the destination and that they have a contribution to make in Australia was evident from in-depth discussions with respondents (undertaken by the author). The following are a representative sample of views on this matter.

> My overseas work experience has been extremely valuable professionally, but I would like to return to Australia to improve my lifestyle and contribute something to the system in which I was raised and educated.

> I love Australia and want to return and hopefully make a contribution to it. I will need to compromise career to do so.

> In my case long absence has led to an exciting opportunity to help Australian high tech firms to better export their products.

> My intention is to transfer my operations back to Melbourne, taking three staff with me. We intend to have operating bases in Singapore and Shanghai.

The role of the returning diaspora as an agent of change in their home country has rarely been examined, although a detailed examination of the pivotal role played by returnees and the diaspora in Greece is instructive in a comparative sense (Minoglou and Bitros 2006). In the Australian context there has been no detailed research on the impact of returnees, to date. Despite this lack or omission, there would seem to be considerable potential for many positive impacts to accrue since: (i) most of the returnees were highly skilled when they left Australia and they have extended their expertise and experience while away as well as developed a network of overseas contacts which can benefit their work on returning home; and (ii) most return at a stage of the lifecycle when they have a considerable working life in front of them. Indeed, as the age-cohort data of returnees demonstrates, they are often at the peak of their careers (and in the family-formation stages of their life courses).

Moreover, the contemporary labour market in Australia is very tight and skill shortages in many sectors are a constraint. Indeed current immigration levels are the highest for three decades. In addition, with the resident population ageing, the 'baby boom bulge' is poised on the cusp of retirement. This means that over a quarter of the workforce will retire in the next 15 years, opening up a large number of opportunities, many of them at senior levels. The potential for returnees to be important agents of change upon their return to Australia is therefore considerable.

From an Australian perspective returning expatriates have a number of advantages over immigrants. Their qualifications are readily recognized and accepted by employers, they have knowledge of local labour markets and context, there are no language barriers and they have established social networks. Yet despite these apparent advantages, there is some evidence that returnees are not experiencing such a smooth transition back into the Australian workplace, as the above observations might suggest. Several respondents in the surveys made reference to the fact that when they return many suffered at the hands of a 'tall poppy' syndrome. This refers to an aspect of the egalitarian ethos in Australia which seeks to diminish the achievements of people who stand out from the crowd (unless it is in sport) and there is accordingly a cutting down among 'tall poppies' that rise above the average. Returnees are sometimes perceived negatively by employers and some expatriates report bitter experiences on return.

> I feel very bitter about not being able to fit in [Australia] so I returned to a country that values my skills and provides me with more opportunity [Japan].

> My brother, myself and two of his best friends were all high achievers and may be a bit more academically adventurous than average Adelaide people. We all separately came to UK and have never returned, finding the scope to grow much greater in UK.
>
> I'm sad about this because Aus is great but nothing I did when working was allowed or valued there and my brother (who sold his business for over $100 million in the US) was unable to get going there either. In Aus. we were regarded as misfits and odd, in the wider world we have been regarded as entrepreneurial genius types. Unfortunately we have felt Aus. has an ethos of 'no tall poppies'.
>
> I am an Australian who spent 15 years in Japan, after a not unsuccessful career in Australia to come back to a very different place. The thing I have found is the 'reverse culture shock' is the toughest thing of all to cope with.
>
> I expected some type of interest in my CV and experience after I came back, and found very little.

In some cases, respondents reported that they had returned to Australia and were not able to fit in. They felt they had experienced difficulty in gaining recognition for their foreign-based achievements and had subsequently left Australia again. As a consequence, there has been a special network set up in Australia to support returning Australians in response to such problems.

There are many stories recounted by returning expatriates of them being discriminated against by potential employers, but also of there being a wider negative attitude toward expatriates by the Australian community. One quite typical case, observed:

> It has been my personal experience that many Australians feel somehow that leaving Australia for long periods is somehow unnatural.

However, with the increasing scale of emigration of young Australians becoming very obvious and real, this 'opinion' would appear to be changing. In a survey of 1,000 resident Australians, Fullilove and Flutter (2004, 38–41) found that 91 per cent agreed with the statement that expatriates are 'adventurous people prepared to try their luck and have a go overseas' and 75 per cent agreed that expatriates are 'doing well for themselves'. Some 44 per cent indicated that the existence of the diaspora was a good thing for Australia, while the proportion agreeing to this sentiment was 62 per cent among young people. Hence it appears that this 'culture of negativity' about expatriates may be breaking down in 21st century Australia.

Return Migration and Circulation's Impacts and Significance

It has been demonstrated here that return migration should no longer be overlooked, ignored or disparaged. On the contrary, it has been proven to be a significant structural element in an Australian migration system, in which the overwhelming focus among policy makers and researchers has been on the in-movement of foreign immigrant settlers, until recently. The long-term circular migration of Australians to and from other countries represents not only a major part of the nation's demography, but has the potential to be an important contribution to the nation's society and economy. The Australians involved are overwhelmingly young adults and they live and work in different economic, social and cultural situations for several years before returning. When they come back they have enhanced skills, contacts and experience which are likely to mean that their economic contribution in Australia is greater than had they remained at home. Moreover, as potential, or actual 'social agents of change' they come back to Australia having seen different ways of doing things, lived in different cultural context and they have been influenced by a more diverse range of influences than they had in Australia while growing up.

The experiences of young Australians who have lived and worked for extended periods overseas and then returned will be of value to Australia as it competes in an increasingly internationalized economy and goes about establishing its rightful place in a rapidly globalizing world. In assessing the nature and scale of this impact, it is necessary to consider not only what they bring back that is different, but also the numbers of young Australians involved. In 2006–2007 some 39,373 Australian residents and citizens aged 20–29 left the country on a long-term basis. This represents only 1.5 per cent of the total population in this age group. However one could argue that young people generally would take only one such long-term move out of Australia while aged in their 20s, so it is probably around 15 per cent of Australians in their 20s who move out of the country for more than a year or permanently during their 20s. Moreover, their impact on the Australian society and economy is likely to be out of proportion to their numbers, since it has been demonstrated that they are strongly concentrated in the highly skilled groups.

Another important impact that the young returning Australians have is related to marriage migration. Most of the long-term and permanent emigrants and the returnees are aged in their 20s and 30s. These are not only the ages at which the highest propensity to migrate occurs, but they are the ages at which Australians establish life partnerships. Hence a significant number of returnees bring with them (or have follow them) a partner they met while living and working in a foreign country. In 2006–2007, 40,430 persons entered Australia under the Spouse, Prospective Marriage (fiancé), Interdependency (mainly same sex partner) Partner Visas (DIAC 2008, 38). While they include partners whose marriage had been arranged through traditional marriage markets (e.g. in the case of Indians) or so-called 'mail order brides' (e.g. in the case of the Philippines) the majority were the results of liaisons made by Australians temporarily living and working in foreign countries. To put this figure in context, in 2006–2007 there were a total of 114,222 marriages celebrated in Australia. The return migration of young Australians is thus contributing significantly to the increasing multicultural diversity in Australia. This is evident in Table 10.15 which shows the origins of the partner visa recipients in 2006–2007. The increasing significance of this phenomenon is the fact that visa entrants to Australia under the partner visa category increased by 51.7 per cent during the enumeration period,1999–2000 to 2006–2007.

Table 10.15 Australia: citizenship of partner immigrant visa recipients, 2006–2007

Country	Per cent	Country	Per cent
United Kingdom	13	China	10
India	8	Philippines	6
Vietnam	6	USA	4
Thailand	4	Lebanon	3
Indonesia	3	Other	3

Source:DIAC 2008, 38 (n=40,437).

Discussion

Australia is, quite clearly, experiencing a significant 'brain circulation', in the sense that it is experiencing a substantial, and increasing, outflow of highly skilled residents and citizens with a majority of them returning, most within a few years. It is important, however, to not view this pattern which has been quantified and analysed here as being 'exceptional' (and unique to Australia, by demonstration). The apparent divergence from conventional wisdom or apparent dramatic difference from other nations' return experiences explained in this chapter is in large part because Australia has the quality and comprehensiveness of data, which has allowed the full scale and composition of both the outflow of its nationals and their return to be measured. Such comprehensive data enumeration and analysis of circulation and of the migration flows of returning nationals is not possible in most countries because they do not collect information on people leaving the country, while many also do not compile data on nationals returning. Accordingly, in the Euro-American dominance of the literature it is not surprising that the significance of return migration has been overlooked. On the other hand, the Australian case must be seen as being indicative of the importance of return migration in other Euro-American countries and of its relevance in the development of their human and social capital. Moreover, there are indications that policy intervention can facilitate this impact.

Return migration is a long-standing (albeit neglected) feature of international migration (King, 1986) but it has taken on a new significance with the paradigm shift in international migration research from permanent emigration/immigration and settlement to transnationalism. This shift has seen a transfer of focus away from the 'historic immigration narrative of departure, arrival and assimilation' (Ley and Kobayashi 2005, 112) toward interactions linking origin and destination. Return migration clearly occupies a central position in a transnationalism approach to migration, whereas it was peripheral in the 'assimilation narrative'. However, while return migration must not be seen as a newly emerging phenomenon associated with the age of globalization, there are elements of it which are new. What are these dimensions which differentiate contemporary patterns of return migration of Australians from the past?

Firstly there can be no doubting that the present pattern of long-term emigration and return is on a scale unprecedented in the past. In Australia there has been a long tradition of the 'overseas working holiday' which was a rite of passage for young adults from elite families, especially women. This goes back to the nineteenth century and was mainly directed to the United Kingdom. However these sojourns were limited targeting the elite and impinged on only a tiny fraction of society. The situation in contemporary Australia is totally different. The overseas sojourn has become a norm for young Australians of all kinds of backgrounds. Taking a 'gap year' between completing high school and beginning tertiary education is one manifestation which has become common.

A second dimension which is different is a shift in the balance of motivations among young Australians travelling overseas for long periods. Originally the emphasis in the 'working holiday' was very much on the holiday with work being a means to finance travel, sightseeing and vacationing. Moreover work often was outside the regular vocation of the mover and concentrated in seasonal and part-time jobs. However among the contemporary young emigrants from Australia career factors loom much larger in their decision to move overseas. While there are strong elements of 'seeing the world' and 'gaining new experiences' there are even stronger motivations to assist career mobility. The internationalization of labour markets has had a significant impact.

A third element of difference relates to the fact that in the contemporary context return migration must be seen as being one part of a more extensive pattern of circular interaction between Australia and foreign countries associated with the mobility of Australians. Australians overseas are now able to interact with their home country with a frequency and intimacy that was unthinkable until recently. This is apparent in Table 10.16 and Table 10.17

which show the level of contact which Australians living overseas have with Australia. Whereas in the early post-World War II decades the prohibitive pricing of international travel and international telephone calls made it necessary to travel and call home only very infrequently, the contemporary situation has been transformed. Most Australians overseas are in almost daily contact with their home nation through telephone calls, emails and daily reading of newspapers and other media. As a result they are able to keep up with events and trends in Australia as much as colleagues in Australia. They are able to maintain social and professional linkages with Australia in a personal and intimate way that reinforces their identity with Australia and their involvement in activities in Australia. In short, return migration needs to be considered as part of a wider process of transnationalism involving regular electronic interaction, frequent visiting and regular reinforcing of intentions to eventually return to Australia. And, a brief numerical example illuminates one aspect of the increased circulation that is accompanying Australian transnationalism's spatial expansion: the number of short-term trips taken out of Australia increased by 54.5 per cent from 3.3 million in 2002-03 to 5.1 million in 2006–2007, thus indicating the rapid increase in the extent to which Australians are visiting other countries.

Table 10.16 Australians Overseas: number of visits to Australia since moving

Number of visits	Australian Emigration Study 2002		US Expatriates Study 2007
	No.	Per cent	Per cent (n=1,581)
None	282	13.6	13.2
1–4	925	44.6	38.1
5–9	440	21.2	48.8
10–19	262	12.6	
20+	164	7.9	
Total	2,072	100.0	100.0

Source: Hugo et al. 2003; Parker forthcoming.

With Australia recently having become a high income, if still-peripheral, national player in the global economy, it is not surprising that there is a substantial outflow of well- educated young Australians seeking advancement and experience in the professional, skilled labour markets of leading cities in the wider global sphere. The increasing scale of this exodus has been significant enough to warrant an Australian Senate 'Inquiry Into Australian Expatriates' in 2003 with the following terms of reference directing the concerns: (i) the extent of the Australian diaspora; (ii) the variety of factors driving more Australians to live overseas; (iii) the costs, benefits and opportunities presented

Table 10.17 Australians living overseas: frequency of contact with Australia, 2006 (per cent)

Frequency of contact	Telephone		Email		Post		Fax	
	Business	Personal	Business	Personal	Business	Personal	Business	Personal
At least once a day	3.5	6.2	8.9	29.8	0.5	0.3	0.5	0.1
Every 2–3 days	3.4	19.3	4.8	28.8	0.8	1.0	0.8	0.3
Weekly	5.6	45.9	7.5	27.9	2.3	4.8	1.5	0.6
Monthly	12.5	21.9	15.5	10.8	8.2	36.5	4.3	3.2
Every 6 months	15.0	4.6	14.5	1.9	11.9	34.7	7.4	9.2
Once per year	9.2	1.1	7.6	0.4	8.4	10.0	7.4	10.5
Never	50.9	1.0	41.3	0.5	68.0	12.7	77.9	76.0

Source:One Million More Survey (n=9,529).

by the phenomenon; (iv) the needs and concerns of overseas Australians; (v) the measures taken by comparable countries to respond to the needs of expatriates; and (vi) ways in which Australia can better use its expatriates to promote economic, social and cultural interests.

It should be noted, however, that there was no direct reference to return migration in this enquiry's mandate. The Senate Committee brought down its report in March 2005 (Australian Senate Legal and Constitutional References Committee, 2005). It made 16 recommendations including the following nine: (i) measures to improve better provision of information to expatriates; (ii) establishing a policy unit on expatriates within DFAT; (iii) improve statistical information on expatriates; (iv) revise consular role of missions to better engage expatriates; (v) improved registration of expatriates in missions; (vi) amend Citizenship Act in a number of ways including to enable children of former Australian citizens to apply for Australian citizenship; (vii) enable some expatriates to remain on electoral enrolment; and (viii) encourage non-profit organizations to pursue philanthropic contributions from expatriate Australians.

Again, no specific reference to the potential role of return migration emerged among these recommendations. Of the 16 specific recommendations (none concerned with return migration) eight were accepted by the Federal government.

The neglect of the return migration issue in this Senate Inquiry is nothing new. While return migration was one of Ravenstein's (1885) original seven 'laws of migration' and a myriad of studies different in time, context and scope have shown that virtually all flows of migrants have a compensating flow (King 1986) in the opposing direction: 'return migration is the great unwritten chapter in the history of migration' (Olesen 2002, 135) In the case of the Senate Inquiry the following factors may have been influential in return migration being overlooked: (i) a lack of data which means that the precise scale of the phenomenon cannot be established; (ii) a lack of research on the scale, characteristics, impact and potential of return migration in Australia; and (iii) a strong philosophy among some migration researchers that emigration is not a significant issue in Australia because of the large scale immigration (Birrell et al. 2001).

I am convinced, however, that an equally strong argument can be made for return migration to be accorded greater attention by Australian policy makers. Australia, like other OECD countries, will have an increasing demand for skilled labour into the future, due to low fertility, ageing and the character of contemporary economic development which places particular premiums on skill acquisition, advanced education and professional experience in the most productive, high return labour markets. Immigration will continue to be a major demographic, economic and social process of considerable influence and importance. At the same time, in an increasingly competitive global market for skilled migrants there would appear to be a strong case for inclusion of

Australian expatriates (and their return) in national immigration policy. The research reported here indicates that there are a considerable proportion of Australian expatriates who have the desire to return to Australia. There is a need for research to establish what is needed to realize this desire among expatriates. While it is apparent from what has been presented that many expatriates in fact do return, it would seem that the potential for return migration is considerably greater than that which actually eventuates. Understanding which factors constrain return migration is an important priority therefore. Australia has a well developed immigration programme designed to attract skilled immigrants to the country. Surely this can be modified to facilitate the return migration of Australians? There is considerable evidence to show that while Australians may be attracted overseas in the early stages of their career, later many are ready to return when they begin family formation. Importantly, they retain a desire for their children to grow up as Australians with access to their extended family

While there is a lack of policy interest in diaspora and return migration in Australian government circles it is also an unfortunate reality that there is little or no research in this area to provide an empirical-base that might better inform program development and policy formulation. This is because the research agenda is still predominantly locked in the paradigm of migration, which dominated in the first five post-World War II decades in Australia and focused exclusively on permanent settlement of immigrants. This has not only been the case in Australia however. Emigration studies more generally are greatly outnumbered by immigration analysis and policy making and usually the focus has been squarely on issues related to permanent settlement at the destination. Emigration and non-permanent migration have not been on the Australian radar screen until recently because of an overwhelming focus on immigration and settlement. However, the emergence of transnationalism as the dominant paradigm in global international migration and the rise of transnational communities, which transcend national borders, make it imperative that a reorientation and widening of scope occurs both in research and policy.

References

Australian Senate Legal and Constitutional References Committee (2006), *They Still Call Australia Home: Inquiry into Australian Expatriates*, Canberra: Department of the Senate, Parliament House.

Azure, A. (2003), 'Leaving and belonging', in Havenhand, B. and MacGregor, A (ed.), *Australian Expats: Stories from Abroad*, Newcastle, Australia: Global Exchange, 27–33.

Bedford, R., Ho, E. and Hugo, G.J. (2003), 'Trans-Tasman migration in context: recent flows of New Zealanders revisited', *People and Place*, 11(4): 53–62.

Birrell, B., Dobson, I.R., Rapson, V. and Smith, T.F. (2001), *Skilled Labour: Gains and Losses*, Canberra: DIMIA.

Conway, D. and Potter, R.B. (2007), 'Caribbean transnational return migrants as agents of change', *Blackwell Geography Compass*, 1(1): 25–45.

Department of Immigration and Citizenship (DIAC), *Immigration Update,* various issues, Canberra: AGPS.

Department of Immigration and Citizenship (DIAC) (2008), *Population Flows: Immigration Aspects*, Canberra: AGPS.

Department of Immigration, Multicultural and Indigenous Affairs (DIMIA), *Australian Immigration: Consolidated Statistics*, various issues, Canberra: AGPS.

Dumont, J. and Lemaitre, G. (2005), *Counting Immigrants and Expatriates in OECD Countries: A New Perspective*, Paris: OECD, http://www.oecd.org/dataoecd/27/5/ 33868740.pdf.

Fullilove, M. and Flutter, C. (2004), 'Diaspora: The World Wide Web of Australians', Lowy Institute Paper No. 4, Lowy Institute for International Policy, New South Wales.

Hugo, G.J. (1994), *The Economic Implications of Emigration from Australia*, Canberra: AGPS.

Hugo, G.J. (2006a), 'An Australian diaspora?', *International Migration*, 44(3): 105–32.

Hugo, G.J. (2006b), 'Developed country diasporas: the Example of Australian expatriates', *Espace-Populations-Societies*, special issue on Diasporas and Metropolis, 1: 181–202.

Hugo, G.J., Rudd, D. and Harris, K. (2003), 'Australia's Diaspora: It's Size, Nature and Policy Implications', CEDA Information Paper No. 80, CEDA, Melbourne.

King, R. (1986), *Return Migration and Regional Economic Problems*, London: Croom Helm.

Ley, D. and Kobayashi, A. (2005), 'Back to Hong Kong: return migration or transnational sojourn?', *Global Networks*, 5(2): 111–27.

Minoglou, I. and Bitros, G. (2006), 'Some innovations and network effects of Greek diaspora and entrepreneurship: 1750–2000', presentation to seminar on *European Diaspora – Exploring Opportunities for its Involvement in Economic Development of Countries in South-East Europe and their Accession to European Union*, FH Joanneum, Graz, 3–4 April.

Olesen, H. (2002), 'Migration, return and development: an institutional perspective', *International Migration*, 40(5): 125–50.

Osborne, D. (2004), 'Analysing traveller movement patterns: stated intentions and subsequent behaviour', *People and Place*, 12(4): 38–41.

Parker, K. (forthcoming), 'Engaging emigrants: experiences of the Australian diaspora in the United States', PhD thesis, Department of Geographical and Environmental Studies, University of Adelaide.

Ravenstein, E.G. (1885), 'The laws of migration', *Journal of the Royal Statistical Society*, 48(2): 167–235.

Rogers, A. (2004), 'A European space for transnationalism', in Jackson, P., Crang, P. and Dwyer, C. (eds), *Transnational Spaces*, London and New York: Routledge, 164–82.

Schachter, J.P. (2006), *Estimation of Emigration from the United States Using International Data Sources*, New York: United Nations.

United Nations (2006), *International Migration 2006*, New York: United Nations.

Wood, F.Q. (2004), 'Beyond brain drain' – mobility, competitiveness and scientific excellence', workshop report prepared by Centre for Higher Education Management and Policy, University of New England, Armidale, Australia.

PART 3
Theoretical Generalizations

Chapter 11
Return of the Next Generations: Transnational Mobilities, Family Demographics and Experiences, Multi-local Spaces

Dennis Conway and Robert B. Potter

The publication in 2005 of our collection entitled *The Experience of Return Migration: Caribbean Perspectives* heralded the emergence of a new agenda in international and transnational migration research; namely, examinations of the experiences and adaptations on return of younger and youthful migrants to their ancestral homes in the global South (Potter et al. 2005). Prior to this publication, there had been a few references to such pre-retirement returns in examinations of international circulation (Byron and Condon 1996; Condon and Ogden 1996; Conway et al. 1990; Ellis et al. 1996), in investigations seeking to link return visiting with return migration (Duval 2002, 2004), or in wider samples of returnees to specific Caribbean origins (Byron 2000; DeSouza, 1998; Duany 2002; Gmelch 1987, 1992; Thomas-Hope 1986, 1999). One of the present authors, Rob Potter, had published work on young returnees to Barbados a few years earlier (Potter 2001a, 2001b, 2005a, 2005b), so that research in this area was underway, but the topic largely remained 'under the radar'. Fortunately, we were able to recruit contributors to our 2005 collection with different islands as their research foci, so that we ended up with a wide regional coverage of Caribbean case studies – Barbados, St Lucia, Grenada, Trinidad and Tobago, Puerto Rico, and Jamaica. Accordingly, comparative insights could be drawn on these pre-retirement, young or youthful return migrants' experiences and adjustments 'back home'.

Common assumptions among several of the earlier (pre-2005) studies on Caribbean return migration, were that many, or most, return moves were being undertaken on retirement by the elderly 'first generation' of post-World War II emigrants, and that professionals' moves 'back home' were usually after short sojourns overseas (King 2000). As special cases, continuing colonial or post-colonial ties that facilitated circulation between the French-Caribbean territories and France, and Puerto Rico and the United States encouraged repetitive visiting, short-term sojourning and circulatory practices, in which return was greatly facilitated by the lack of national borders (Condon and

Ogden 1996; Duany 2002). In short, as Ghosh (2002: 1) remarked more generally:

> Although an integral part of the migration process, return movement, including its social land economic implications, has so far remained inadequately unraveled in the migration debate. One of the most-neglected areas of migration research, it also has failed to receive adequate and systematic attention from policy makers.

Thus, it was at a noteworthy juncture that our 2005 *Experience of Return Migration* research anthology, together with a 2006 collection by Plaza and Henry entitled *Returning to the Source*, were able to demonstrate that in today's globalizing-and-transnational world, return migrants, and those repetitive circulators who are undertaking more temporary sojourns or visits, are no longer insignificant demographic cohorts. Both collections provided wide coverage of the Caribbean region, with studies being conducted in island nations, among overseas communities, or including both as the contextual frameworks. Regional specialists on migration and return contributed to both collections, and the two generally complimented each other in their coverage.

Generalizations from this Collection's Caribbean Predecessor: *The Experience of Return Migration*

The findings of our 2005 collection were often reinforced by those of Plaza and Henry's 2006 companion, But as a predecessor to this current volume on *Return Migration of the Next Generations*, here we specifically focus on our earlier volume in order to draw out comparative generalizations about the regional dynamics of 'transnationalism and return' that could be pursued further. Several common generalizations about these new waves of Caribbean 'citizens by descent' or 'returning nationals' who are currently 'coming back home' were upheld. No longer consisting mainly of returning retirees, today's Caribbean return cohorts were becoming noticeably more diverse, with respect to age and family-life cycle characteristics and class, social and gendered positions. Their transnational family networks, migration histories and the multi-local nature of national diasporas also brought more diversity to the global-to-local interactions that helped mould both their changing societies and their changing lives. And, despite their numerically small proportions, many were demonstrating that they could act as influential 'agents of change' (Conway and Potter 2007; Gmelch 2006; Potter and Conway 2008).

In addition, the importance of 'pull' factors in promoting return migration was stressed, though assimilation and adaptation problems in 'the other man's world' were not absent, by any means. Negative experiences in the metropolitan societies from which they were returning, such as being made to feel like second-class citizens, the occurrence of racial discrimination and harassment, feelings of anomie and alienation brought on by unfortunate experiences, economic hardship, social disquiet, lack of familial support, and such, were also part of many returnee's mix of reasons. But, undoubtedly it was the pull factors of the Caribbean region, including the climate, returning to family roots, and the availability of opportunities – albeit selective and promissory – that seemed to be of vital significance in promoting return. At the same time, the chance for returnees to improve their standards of living and overall quality of life – for themselves and their children – was always likely to be central to their return calculus.

Several of the studies showed that a prime motive among these cohorts of young and youthful returnees was the feeling, or conviction, that the Caribbean is the best, or better, place to bring up children. And, as their 'first generation' parent or parents has/have often returned on retirement, the core of an inter-generational family was also on hand to help with the care of children and to provide family support networks in times of need. Others talked of the need to return in order to fulfil family duties, such as providing care for these ageing parents. In more than a few instances, the agency of national pride and wanting to do something for the 'motherland' was revealed as a motivational factor prompting return migration (Potter et al. 2005).

The chapters in *The Experience of Return Migration* demonstrated quite convincingly that return migrant experiences differed according to island context, and the levels of development, urbanization and economic diversity that each territory had reached, or was moving towards. 'Changes' were not so much viewed as challenges to these returnees, but rather they offered more diverse employment opportunities, improved living standards and more material benefits, while societal continuities also persisted in terms of a better quality of life, of care-giving environments for child-nurturing and raising, and strong senses of place and national pride. In addition, the metropolitan backgrounds of returnees, and the extent to which returnees were well-supported by transnational networks of extended and nuclear families also influenced the ease of return adjustments for many.

On the other hand, the contributions to this 2005 volume suggested that return migrants of whatever age or background faced a range of common adjustments and ultimately, in some instances, they evinced firmly-held frustrations in coming to terms with their new island homes, especially in their work places. For women in particular, 'competition for men' emerged as an interesting aspect of social distancing between returnees and indigenous stay-at-homes. In several cases, this was also shown to be closely linked to returning women's difficulties in making female friends. In extreme instances, this appeared to have led to a degree of 'othering', that is the manifest marginalizing of the returnees as outsiders who are fundamentally different from the resident, national population.

More common was the generalization that a sizeable number of young returnees were characterized by their 'betwixt and between-ness'; that is, their transnational existence and identification with two or more 'worlds within worlds' – French, British, Canadian, American and Caribbean, black and white, 'hybrid' and 'mixed-up'. Some returnees also felt frustrations at the poorer facilities, limited shopping opportunities and the higher prices that they experienced in their new Caribbean homelands, in comparison to the metropolitan areas they left behind in Europe, the UK or North America. In other cases, power cuts, water shortages and domestic tasks being harder to complete were day-to-day irritants that made adaptation on return difficult, but not insurmountable.

Most saliently, as well as the expected counter-stream of returning first generation retirees, more Caribbean youthful and younger migrants of working age, in their 30s and 40s, had decided to give it a try 'back home' (Conway and Potter 2007; Potter and Conway 2008). These young returning nationals, however, had more information available at their fingertips via the internet, telephone and other global networked systems, and as a direct consequence, were more directly aware of the opportunities that were open to them, and also perhaps to the adjustments they were having to make in their new environments. They were likely to be more skilled than their migratory counterparts were in the past, in large part because they had availed themselves of educational opportunities while abroad; either as their birth-right as second-generation overseas-born, or as a post-colonial-right that their 'mother country' owed them. Some might very well be moving back and forth, trying out strategies to see if they could enhance their standards of living and the quality of their lives as they passed through their family-formation and career-formation life-course stages.

We were able to conclude, therefore, that many Caribbean return migrants inevitably experienced frustration while they adapted, whilst others adapted more easily and quickly. Others retained ambivalent and conflicting views of their position 'betwixt and between', and their multiple identities either helped or hindered their adaptation experiences. By definition, those who returned and remained were the ones who had developed niches from which they could build island contacts, make new circles of friends, and generally participate in the social and economic fields associated with their professions and businesses in the island homes of their parents – now their homes and nations. Where this happened, dual citizenship and multiple, transnational identities were becoming commonplace, rather that the exception. As transnational migrants, with overseas experience and a pragmatic flexibility in their approach to life in the present and future, returnees in their 30s, 40s and 50s were choosing island life, and had decided to live, work and play in the Caribbean island 'homes' of their ancestors, rather than struggle to make their livelihoods in the metropolitan centres they left. Many did not, however, sever their ties with the metropolitan centres of their recent past, but rather kept in close touch with family and friends there. Some held on to property, maintained bank accounts and by and large, adhered to transnational strategies to live in and between two worlds, or sometimes three or more, if their family's international reach was multi-local. Some in the private sector, maintained transnational business ties, or began to build transnational businesses as 'tropical capitalists' (Portes and Guarnizo 1991), with cross-border business incubation, expansion and innovation-transfer being an integral part of their newly re-configured multi-local, entrepreneurial environments.

Onward to the Successor: *The Return Migration of the Next Generations*

It is quite clear from the above summary, that there were many generalizations we could derive from our first Caribbean comparative project that offered a fruitful path to follow. The genesis of the idea to widening the scope beyond the aforementioned regional focus and conduct a more global comparison came while Rob Potter spent a month-long 'busman's holiday' with Dennis Conway in Bloomington as a guest scholar of Indiana University's Institute of Advanced Study in November 2006. Dennis had undertaken Caribbean-Pacific Island comparative work on remittances investment practices and outcomes with John Connell (Connell and Conway 2000), so that was a logical extension to pursue and widen. Transnationalism and return migration scholarship in the wider Asian-Pacific region was another domain to tap, and the resultant co-operation of our far-flung contributors was readily and enthusiastically cemented via the internet. Several were renowned scholars with decades of research experience, while a few were in the early stages of their academic careers, so the mix of contributors – geographers and anthropologists – was a suitable balance of disciplinary reach, while all had impressive records of field experience. The final step was to bring in Ashgate, which had published the earlier 2005 work, and convince them of the growing global significance of these youthful, *next generations*.

Discussion

Following an introductory chapter, the collection is divided into two sections; the first comparing second-generation return migrant experiences in five chapters, the second comparing young and youthful return migrant experiences more widely in four chapters, with the present concluding chapter building more global comparisons as logical extensions, qualifiers, or modifiers, of our collection's recent findings from the insular Caribbean region. The geographical mix is considerable. Three chapters (2, 3 and 8) deal with return experiences in Pacific island territories and their metropolitan neighbours New Zealand and Australia. In Chapter 2, the MacPhersons –

Cluny and La'avasa – find considerable diversity in the reasoning behind the return of adult Samoans, most of whom are 1.5-generation returnees who had lived much of their lives abroad in New Zealand. In Chapter 3, Helen Lee focuses her lens upon young second-generation Tongans, mostly between 18–30 years of age, who have lived in Australia before returning, or are still living in Australia (and were interviewed there). In addition, Lee talked to a small sample of teenage Tongans, or 'adolescent returnees', who, along with 'young deportees' she characterized as 'forced return migrants'. In Chapter 8, John Connell first sets the stage by taking a comprehensive look at the return of health professionals – nurses and doctors – to the South Pacific, and then narrows in on Niue and the Cook Islands to examine the particular patterns of health worker return in those small islands.

Two chapters (5 and 9) examine transnationalism and return in Caribbean islands and their overseas metropolitan 'social fields' in the UK and North America. In Chapter 5, Rob Potter and Joan Phillips review the 'Bajan-Brit' return phenomenon, the return of second-generation women and men from the UK to Barbados – the island home of their parents. In Chapter 9, Trinidadian and Tobagonian 'prolonged sojourners' are the group of transnational returnees that Dennis Conway et al. interviewed, and it is their 'narratives' that tell the stories of transnational experiences and practices of this predominantly middle-class transnational cohort.

The remaining case studies are individual chapters representing a wide geographical reach into areas of the global North and global South. In Chapter 4, Eunice Akemi Ishikawa details the case of Japanese-Brazilian returnee families and their second- and third-generation children's experiences on return to Japan. In Chapter 6, Anastasia Christou deals with second-generation Greek-American return to Greece. Chapter 7, by David Ley and Audrey Kobyashi examines the return in mid-careers of Canadian-Hong Kong transnational sojourners, most being educated professionals and members of the highly-skilled global labour force that appears comfortable being 'workers without frontiers'. Finally, in Chapter 10 Graeme Hugo illustrates the exceptional case that Australia provides and offers as a model for the future, by demonstrating that large numbers of circulating nationals can be prompted to emigrate, then return, and even re-return later, if the welcoming environments and encouragement mechanism for migration and return by host and sending country-partnerships are efficiently and effectively managed by enlightened public policy-making.

Geography Matters

One generalization that becomes apparent from this collection's examination of second-generation experiences on return, and of the return experiences of youthful sojourners or circulators, is the importance of the geographical and socio-historical contexts that frame the migrations; whether they are emigrations, circulations and return migrations. As Doreen Massey (1984) so convincingly observed 'geography matters', and nowhere is this more apparent than in the comparative experiences of migration and development relationships in the South Pacific, Polynesian region – the islands of Samoa, Tonga, Niue and the Cook Islands, and the insular Caribbean experiences – in Barbados and Trinidad and Tobago. All were colonial dependencies, but while the latter two Caribbean territories have morphed into post-colonial, politically independent societies, and have enjoyed quite impressive societal growth and economic transformations in contemporary times, the South Pacific experience has not been as positive, or transformative. Elsewhere, John Connell (2003) has been especially sharp in his critical depiction of the South Pacific as 'an ocean of discontent' wherein contemporary migration of the region's island peoples has largely been in response to deprivation and underdevelopment. 'MIRAB-dependency' (Bertram and Watters 1985; Connell and Conway 2000) still plagues the far-flung small island territories of the South Pacific.

Although retention of formal colonial ties to New Zealand facilitates transnational mobility between Niue and the Cook Islands, these institutionalized, dependent relations have not brought about the transformations of island economies, or island societies that such mobility might be expected to help via remittances and return migration. Rather, the comparisons and comparative experiences appear to severely problematize the return of health workers from Australia and New Zealand, who are unwilling to return to the underfunded, work-stressed and technologically-limited, public health sector they voluntarily left behind to pursue career advancement and training abroad, as Connell observes in Chapter 8.

Further, as Lee points out in Chapter 3, Tongan youth growing up in Australia appear to voice the same objections to returning to this underdeveloped, island homeland of their parents; with assimilationist views being paramount among this youthful, teenage cohort. The forced nature of the return of most of these dependent youth, whose Samoan parents dictated such repatriation, has quite obviously not endeared these 'family-deportees' to life 'back home'. It turns out that among the second-generation and even the 1.5-generation of Tongans born and raised in Australia, or brought over as children, ties to their ancestral home were never as strong as those of their parents'. 'Ambivalence' characterized their attitudes and emotional ties to Tonga, so that even when they were partially convinced that such a search for their ancestral cultural identity was justified, and did go back, many experienced problems of adjustment on return and planned to re-return, rather than stay 'for good'. Australia and New Zealand's superior health, education and welfare systems do, quite convincingly, promote re-evaluations of the ties to island homelands among the youthful second- and 1.5-generations, without completely severing them, or causing return to be postponed indefinitely. If a return to Tonga is planned, however, it is more likely to be viewed in temporary terms, as a sojourn during a stage in life, where re-connection is most beneficial, when parental obligations or familial obligations of care-giving are best practiced, or when 'bonding' and familial responsibilities bring social returns to the mover, without restraining, or limiting the options for a re-return, or for a future move on to another overseas opportunity.

By comparison, the contexts of Samoa and the two Caribbean islands in the collection are much more favourable for returnees in terms of societal advances, the growth and development of public health, education and welfare services, and private sector vibrancy in their respective domestic economies. The Macphersons in Chapter 2 note that well-qualified Samoans, who are mostly 1.5 generation, are returning to their ancestral home from New Zealand for various reasons and on varying missions, but some are indeed, professionals and entrepreneurs, as well as returning nationals seeking to deepen their cultural understanding of their homeland, fulfilling family obligations, 'doing good' as social idealists, and returning to make a difference. They are returning to a different Samoa than the one they and their parents left as children. In Barbados (Chapter 5), returnees are finding considerable economic diversification in a mature tourism industry, in producer services, off-shore financial services and the island's well-developed domestic commercial, insurance and retailing sectors. Yet, workplace adaptation is still a problem for some, Barbados' tight social circles take some time to break into, racial affinities take some time getting used to, but many find their transnational status privileges them in comparison to residents'. In Trinidad and Tobago Chapter 9) there is much more industrial diversity, with the mature industrial systems of petroleum and natural gas extraction and servicing, manufacturing ensembles of iron and steel products, methanol and nitrogenous fertilizer refineries, and food processing, cement, beverages, cotton textiles production for domestic and Caribbean markets, providing openings for these incoming skilled workers. In addition, there is a growing 'alternative' and mass tourism industry (in Trinidad and Tobago, respectively), Carnival-related creativity, architectural, art and media sector expansions that are attracting returnees, along with private corporate sector expansion in communication, financial and insurance fields as well. Both of these latter Caribbean 'middle-income countries' (because that is what they have become according to the World Bank) offer numerous opportunities for skilled professionals to make and advance their careers back home, so it is not surprising that mid-career shifts are occurring among the transnational middle classes of both of these Caribbean island nations.

Other case study comparisons in our collection either add to one of these opposing viewpoints, or provide more geographical and social variety to the global mix. Ley and Kobayashi's examination of the often well-planned, life-course pathways of a cohort of transnational sojourners moving between Vancouver, Canada and Hong Kong in Chapter 7 finds resonance and reinforcement in Conway et al.'s Trinidad and Tobago study detailed in

Chapter 9, relating to a sample of prolonged sojourners who have returned after more than 12 years abroad (though the average length they stayed away is 17 years). Of course, Hong Kong is much more advanced and economically dynamic than Trinidad and Tobago, being a major Chinese and global finance centre and a 'first tier' global city in every sense. Indeed, the opposite transnational comparison between Canada and Hong Kong is thrown into relief in this examination, because North America/global North Canada emerges as the slower-paced environment, with a quality of life pace and style that attracts these peripatetic transnationals, while the comparatively lower wage regime they can access in Canada (even as professionals) invariably disappoints them, and encourages them to return to greener pastures in Hong Kong. East Asia/global South Hong Kong, on the other hand, offers these transnationals economic and career advancement, albeit allied with a hectic and fast-paced life. Hence, retirement or re-return to Canada is planned to take advantage of the quality of life and slower pace, the benefits accruing to children, and to avail themselves and their dependents of the accompanying substantive welfare, education and health systems that are comparatively less expensive than in Hong Kong.

Hugo's in-depth analysis in Chapter 10 of Australia's emigrant and return migrant sub-populations is exceptional on several counts. Firstly, the quantity and quality of the data enables him to characterize these cohorts' demographic, socio-economic and migration patterns. Secondly, the overall youthfulness of the circular flows that the country has been experiencing as a sending society firmly counters the notion that return is predominantly a retirement phenomenon. Thirdly, and possibly because it is still a peripheral and emerging nation, her young Australians are moving away to acquire higher education, professional experience and on-the-job training in the 'welcoming' skilled-labour markets of contemporary global cities. Fourthly and finally, though their duration of absence is often lengthier than initially anticipated, and return may be postponed beyond the intended 'duration of stay', many in the Australian diaspora are eventually returning, and those staying abroad still retain the wish that one day they (and their immediate family) will return 'home'. Both the Hong-Kong-Canadian case (Chapter 7) and the Trinidad and Tobago case (Chapter 9), also identify circulation, temporary sojourning and temporary return as similar rational migration behaviours to Hugo's Australian 'brain circulations', though their geographical contexts differ markedly, and the Australian case is demonstrably, exceptionally large in volume, as well as of immediate policy-significance in terms of its national impacts.

Another exceptional case, though differing substantially from all others, is Ishikawa's examination of the experiences of second- and third-generation returning Japanese-Brazilians in Chapter 4, which has been briefly referred to earlier in this 'Discussion' section. There is, of course, the extremely wide 'cultural chasm' between Japan and Brazil, that perhaps is enough to warrant this study's geographical exceptionalism. In addition, however,

Ishikawa specifically focuses upon Japanese-Brazilian school-age children, their educational and assimilation problems, and the decisions that their parents make about their own work regimes and their children's upbringing, as temporary sojourning labourers, albeit legally employed. Thrown into high relief in contrast to many of the other contributions in the collection, are the institutional and legal restrictions surrounding these return migrants' recruitment back to Japan to help meet the domestic labour shortages in manufacturing, while their integration into the 'welcoming' nation is not at all anticipated either by the migrants who retain intentions to re-return to Brazil, or by the Japanese host society. Considerable restrictions abound, reducing these Japanese-Brazilians to second-class status, to temporary contract workers, to home renters (not home owners), and to alien quasi-citizens. Little wonder, that these Japanese-Brazilian second- and third-generation's experiences are fraught with tensions, cultural dissonance, categorizations of them as 'outsiders' and a general unwillingness on the part of the resident population to treat these compatriot, returning nationals as socially equal, or socially acceptable, and to be granted full assimilation.

Christou's micro-scale study in Chapter 6 of two young female, middle-class, highly educated and highly skilled Greek-Americans who return to Athens, Greece, their parent's home, revolves around an in-depth analysis of their life story narratives, which were recorded in 'reflective journals' as an extension of earlier semi-structured interviews the author had conducted with these primary informants. Christou is not only interested in how these returnees construct and implement their 'return project', but with their encounters with antagonism and exclusion, their ambiguous sense of belonging and the ethnic and cultural complexity of their 'trans-cultural selves' and the ways they struggle with negotiations over their dual identities on return to their ancestral homeland. The Athens, to which these second-generation Greek-Americans returned was, however, not the same city and capital of the ancestral homeland that their Greek parents had left for America. It had undergone a transformation, was an immigrant city, and was much more cosmopolitan, modern, and metropolitan in social mix and had enjoyed considerable social reordering, structural modernization and enlargement and had benefitted from global-to-local linkages, in terms of its standing in the eastern Mediterranean region (Christou and King 2006). The narratives of these two returnees are viewed/translated as 'journeys of self-discovery' in which ambiguity and self-reflection predominate in their articulations of 'feelings and meanings of being, becoming and belonging'.

Age and Life-course Stage Matters

Another generalization of note is the considerable differences in experiences of returning migrants in terms of their ages, both when they are interviewed and when they are undertaking their return migrations. Age and life-course stage turn out to be critical demographic and social markers that influence the timing of emigrations, returns, temporary circulations, as well as sojourning period lengths, decisions to reconsider migration, re-return, and such. In several of our contributions, the evidence is clear that the ages and life course stages at which emigration is enacted, when return visits are repetitively made, when return migration is decided upon, and when re-return is planned, and/or executed, are all important aspects of the mobility decision-making. For children in their nurturing and 'growing-up' phases, parental responsibilities and decisions usually dictate their mobility, but in the later stages of transition to adulthood, education-led international migration commonly occurs. Individual mobility can then morph into partnered or tied-movement with age and family-formation considerations likely to follow, with the life courses of emigrants and circulators being instrumental in the timing and planning of more mobility, of the lengths of sojourn, of postponements to return intentions and extensions of the lengths of stay, and so on.

Substantiating these generalizations, two cases in our collection drive home the point extremely effectively. Thus, in Chapter 10, Hugo's Australian data are particularly convincing in that as many as 15 per cent of young Australians undertake a long term move out of Australia while in their 20s. The bulk of Australian return migration is of people in the 25–39 year old age bracket, so they are still relatively young and in 'mid-career' when they undertake the return circuit. An additional wrinkle to this circular process of young Australians concerns marriage. As Hugo observes:

> ... a significant number of returnees bring with them (or have follow them) a partner met while they are living and working in a foreign country... The return migration of young Australians is thus contributing significantly to the increasing multicultural diversity in Australia ... The increasing significance of this phenomenon is the fact that visa entrants to Australia under the partner visa category increased by 51.7 per cent during the enumeration period, 1999–2000 to 2006–2007.

In Chapter 7, Ley and Kobayashi's account of the meticulous planning undertaken by some transnational Canadian-Hong Kong dual-citizens they interviewed, adds another 'story' to the substantiation of the importance of the life course stage through which transnational migrants are passing, or anticipate following, in the near future. As they remark: 'transnationalism invokes a travel plan that is continuous, not finite.' Transnational migrants at different stages of their life cycle undertake 'strategic switching between an economic pole on Hong Kong and a quality of life pole in Canada', so that they plan to make the best of both worlds, despite the longer range movement and more costly trans-Pacific air journey of 12–13 hours between Hong Kong and Vancouver (the nearest metropolitan landfall in Canada for them). Some of these transnational 'astronaut' families (Ho 2002) plan to re-return to Canada on retirement, seeking a better quality of life and the security it provides in terms of its public health and welfare systems. Others plan the re-return of the family or their elder children to take advantage of the better (and decidedly cheaper) higher education opportunities in Canada as compared to Hong Kong. Age transitions and life course, or life style, stages are also found to be important considerations of returning or circulating families in other contexts that are under scrutiny in this collection – Barbados, Trinidad and Tobago and Samoa, for example.

Family Matters

The family obligations of children's responsibilities, or of 'bonding' to their parents (Connell in Chapter 8 refers to this time-honoured family tie) were found to be highly influential in many cases of young nationals' return home. The immediate family was also extremely important when a parent or parents made decisions to return, to sojourn, to re-return or to move on. Family 'quality of life' issues, or lifestyle concerns for raising children in a caring environment, were often cited as counters to economic and career concerns, or even as alternatives, where successful careers were interrupted in mid-stage and family concerns superseded such economic calculations to prompt a return home. And, this appeared to be especially significant with the Canadian-Hong Kong 'transnational sojourners' who Ley and Kobayshi interviewed in Chapter 7, in which the planning of sojourns was quite often carefully synchronized with the educational stages their children were passing through. 'First generation' retiring parents' influences also played a part in the return decisions of the next generations, either as direct influences by Trinidadian mothers on their sons and, or as convenient *in situ* 'care-givers' of their next generation's offspring daughters (as documented in Conway et al.'s Chapter 9), or as elderly dependents who needed care given to them by the return of second-generation youth (as documented in the South Pacific cases of Samoa and Tonga (the MacPhersons' Chapter 2 and Lee's Chapter 3, respectively).

Parents and grandparents also get left behind, as their youthful dependents go abroad for higher education, go away to build their own separate lives and livelihoods, but some – as the first generation – take their children with them. The latter family members then become the 1.5 generation of migrants, as they grow up abroad and assimilate, as they are encouraged to retain ties to the homeland of their parents and grandparents, by these 'still-influential' family authorities. 'To keep them in touch', parents may have their children accompany them on return visits to the ancestral homeland, or take independent visits. Grandparents, as first generation emigrants may even be instrumental in sponsoring their third-generations' return visiting to make them better aware of their family and national heritages.

Parents and grandparents may on the other hand, return during the pre-retirement stages of their life course, or on retirement, and such transnational location and relocation, will only further influence (and complicate) their children's options and their young family member's transnational itineraries. Emigrating parents and grandparents may stay abroad, of course, while their children return, move on elsewhere, or stay close by. Family members, whether parents, grandparents or children may thereby distance themselves at least in physical terms, though rarely in terms of transnational communication and contact, unless unhappy, or irreconcilable, separations have occurred.

As pointed out in the Samoan and Tongan cases in Chapters 2 and 3, transnational family obligations influence dependent children's mobility options, so that they may be 'encouraged' to return to the ancestral home to attend to ageing, infirm grandparents, or 'forced' to return because of their emigrant parents' concerns for their secondary education in New Zealand. In other islands of the South Pacific, these youthful expatriates may be 'obligated' through familial bonding to return as health practitioners to attend to their grandparent's welfare as Connell observes in Chapter 8. So, intra-generational family connections feature strongly in many return calculations, whether in the Caribbean, the South Pacific, Australia, or Ireland (see Laoire 2008).

Family love, family obligations and responsibilities feature as strong, enduring factors in determining, or conditioning transnational 'mobilities'- whether they are temporary, circulatory and relatively permanent in nature – and the extended transnational family networks that have become multi-local in their breadth and global reach appear to have become an indelible feature of the contemporary societal landscapes of many originally sending territories of the global South as accompaniments to globalization. As argued elsewhere by one of the authors (Conway 2007), and re-iterated even more convincingly by Mary Chamberlain in her latest 2006 text *Family Love in the Diaspora*, the transnational African-Caribbean families of Britain and the Caribbean (and their North American extensions too) have a lot to offer in terms of their durability, practicality, strength and vibrancy. They very well may be a harbinger, or model of the directions in which contemporary 'multicultural' European and Western families are now heading.

If this can be posited as a tentative conclusion to this section, Mary Chamberlain has some final thoughts that are well worth rehearsing here. She is convinced that there is a pride in the Caribbean *creole* family that revolves around an emphasis on this social unit's 'closeness', and its importance for both support and solace, identity and differentiation. As she so succinctly explains:

> It is not just that you come from Jamaica (or Trinidad or Barbados) but that you come from a particular Jamaican (Trinidadian, Barbadian) *family* that stands for identifiable beliefs and values, and that represents a formidable network of kin whose loyalty – as members of a shared lineage – can be taken on trust. (Chamberlain 2006: 110)

Commenting further on the transnational family's pragmatic and symbolic strengths, Chamberlain (2006: 111–12) offers this summary:

> With a family dynamic and a family culture built on transnational links, and close family networks that stretch around the world, family members are provided with opportunities to utilize these networks to further their own employment or career profile, to broaden their experience base, and to strengthen family resources. The links and networks are both multinational as well as transnational. *Indeed, the spreading or dispersal of material and emotional resources throughout the diasporic trajectory of each family provides diversity and security, strength and opportunity.* Family members provide important points of contact, facilitating migration, re-migration, and of course, return. *When the family is both the source of belonging and the resource of survival, then identity is both portable and secure.*[1]

Circulation and Temporary Return of Growing Importance

With the global recruitment of skilled workers growing apace, and with little indication that global migration will not keep increasing in this post-1980s era of globalization and the 'new age of migration' that is with us (Castles and Miller 2003), it is not so much whether temporary circulation flows will equal or outnumber permanent international migration, but rather that both forms are likely to continue to increase in the foreseeable future (Kapur and McHale 2005; Papademetriou 2003).

Several chapters in this collection point to the temporary nature of return among informants, or samples of returning young and youthful professionals and skilled workers. In the wider literature, other scholars have pointed to the increased social, political and economic returns in social capital that international/transnational skilled migrants who have circulated bring about (Hugo 2003; Lidgard and Gibson 2002; Vertovec 2007). 'Brain circulation' or 'brain exchange' are terms coined to represent this revised opinion on how to counter the 'brain drain' that many sending societies in the global South have experienced (Dawson 2007; Hugo Chapter 10). Temporary migration, repetitive visiting, short-to-lengthy sojourns, and the maintenance of flexible strategies towards transnational mobility and circulation are becoming as well known as permanent emigration and settlement in today's globalizing world of 'cross-border' contract employment, business entrepreneurialism, international tourism's pluralism, transnational family reunification, linkages and obligations, and such.

Does transnationalism 'privilege' such flexible movers? We believe it does, and there is circumstantial evidence in this collection that would point to this reversal of thought, in which the long-held beliefs that international migration

---

1 Italics used to emphasize points in narrative excerpts are the authors' inserts.

'distanced' family members, that it fundamentally replaced one's citizenry obligations and nationalistic identities of one home land for the chosen new one, and that emigration and immigration are no longer predictable, final, or definite transfers of allegiance. Rather, flexibility and fluidity, unpredictability and opportunism, and 'strategic flexibility' are the operational, pragmatic 'watchwords' governing these 21st century migrants' lives and life chances (Carnegie 1982; Conway 2007).

Wider Diasporas, Transnational Experiences and Practices, and Wider Network Reaches are Strengthening Global Interactions, and Helping Build Valuable Human and Social Capital Stocks

The contributions in this collection have examined the return experiences of young and youthful, second- and one-and-a half (1.5)-generations, circulating global skilled workers and professionals and new cohorts of prolonged sojourners. In particular, we have focused on these particular waves of return migrants because they are in the relatively youthful mid-career and mid-family formation stages of their life course. Their circulations and/or returns are undertaken while they are still young, have acquired higher education and professional skills and experiences and accumulated considerable stocks of human and social capital. They have, therefore, plenty of time to 'make a difference', plenty of time to chart their career – and family – futures, and to consider further moves as flexible continuations of their already-experienced, transnational lives and livelihoods, or to decide to stay for good.[2] In comparison to their more elderly counter-part returnees – the first-generation, their parents who may have preceded them abroad, or taken them – these younger and youthful Next Generations have rarely been the focus of migration researchers, until recently (Conway and Potter 2007; Iredale et al. 2003; King and Christou 2008; Potter 2005; Potter et al. 2005; Thomas-Hope 2002, 2006).

Many among the youthful cohorts of return migrants whose experiences our contributors have examined also qualify as global, professional skilled workers, who have been more generally characterized as *Workers without Frontiers* (Stalker 2000), as 'skilled transients' (Findlay 1995), or as 'transnational sojourners' (Ley and Kobayashi 2005). Indeed, the global recruitment for talent and highly-skill professionals in many social and

2 Since, these youthful returnees are being 'intercepted' in the middle of their life course, any decisions about staying put, re-returning or moving on, can be changed as circumstances warrant. Indeed, the transnational experiences they have had abroad, off the island, or 'over there', are likely to be influential in their flexibility towards never ruling out future moves, or future changes of plans to stay longer or permanently, if moving or staying is in the best interests of themselves, their partners and their dependent children, or because of obligations to their dependent parents and grandparents.

economic sectors, in research, technology and knowledge sectors, private and public service sectors, that is underway in this first decade of the 21st century, also involves assessment of returning nationals' potential, and governmental interest in how best to attract home 'the brightest and the best' (Kapur and McHale 2005).

All are examined in a variety of societal contexts; many in the home societies of their first-generation parents, some in metropolitan destinations, but all in 'transnational spaces', in terms of their life-worlds. Many are globally 'in-between', as their multiple identities and transnational circuits of mobility both make them flexible as well as bi- or multinational in affiliation. Many reflect on the ambiguity their bi-national, or multi-national identities present in their interactions with 'others' in the multi-local spaces they themselves feel quite comfortable in, so that surprise, astonishment, even criticism is often expressed at the societal reactionaries they come across 'over there', or on return to their ancestral homeland. For some, their experiences of adjustment are difficult, their high expectations are dashed because of gender, class and ethnic conflicts, and they plan to re-return, or more flexibly and indeterminately do not plan to stay for good. For others, they compromise, they adjust, and the longer they stay on return the smoother the transition becomes. Societal adjustment appears to be easier accomplished than satisfaction in the work place, in large part because family networks play influential and positive roles in social and communal interactions, and family and friends both overseas and at home are strong reinforcing, transnational support systems which are highly valued and firmly held. Dissatisfaction or disenchantment with workplace environments is certainly common among these young and youthful returning nationals, though the level of dissatisfaction does vary according to the territorial contexts of the transnational 'moorings' our migrants view as their spaces. Health workers in particular appear as a singularly 'difficult to please' group of returnees in some 'home spaces', but adaptation experiences in different as well as difficult work environments for many of these overseas trained returnees are to be expected given the wide socio-cultural and economic distances they are attempting to 'bridge' – from Brazil to Japan, from Canada to Hong Kong and back, from Tonga (or the UK) to Australia and back, from America to Trinidad and Tobago, for example.

The signs are indeed very promising, that these young and youthful transnational migrant cohorts will continue to return in appreciable numbers, though they will always be a minority, and never a majority. There will always be a range of adaptation experiences among these returnees, some highly positive, some ambiguous and ambivalent, some disappointingly negative and frustrating. Compromise is apparent among those who find and feel opposite forces repelling and attracting them as they adjust to life 'back home'. Like many emigrants and immigrants before them, returning *next generation* parents may compromise their own careers, hopes and dreams for their children's (the succeeding, third and next generation), so that subjective reality becomes the guiding premise for making the best of things, not an individual objective, rationalization of their individual, or joint, cost and benefit appraisals.

There is promise, as evidenced in some of the cases examined in this collection that the *Next Generations* of young returning professionals will be more effective agents of change than their returning parents, the first-generation retirees. In large part, this is because the overseas experiences these youthful cohorts of returnees have accumulated while in the global North represent much needed social, economic and cultural capital in the developing and modernizing home societies of the global South to which they have returned permanently, temporarily or as a 'brain circulation'. Their transnational experiences have increased their social worth, as flexible, productive and progressive skilled workers, and as committed and purposeful nationals as well as transnational practitioners. Most importantly, many, if not all of these youthful skilled and professional returnees uphold a firmly-held intention to return to 'make a difference', and a firmly-held intention not to turn their backs upon the homeland societies of their birth, or that of their parents. It definitely looks as if significant cohorts of these *next generations* will be returning, so it behooves the home-lands to look for ways in which these young returnees' good intentions can be more than matched by effective welcoming societies, so that there are mutual benefits for all those involved.

References

Bertram, I.G. and Watters, R.F. (1985), 'The MIRAB economy in South Pacific microstates', *Pacific Viewpoint*, 26(3): 497–520.

Byron, M. (2000), 'Return migration to the eastern Caribbean: comparative experiences and policy implications', *Social and Economic Studies*, 47: 155–88.

Carnegie, C.V. (1982), 'Strategic flexibility in the West Indies', *Caribbean Review*, 11(1), 11–13: 54.

Castles, S. and Miller, M.J. (2003), *The Age of Migration: International Population Movements in the Modern World*, 3rd edn, New York: Guilford.

Chamberlain, M. (1995), 'Family narratives and migration dynamics: Barbadians in Britain', *Nieuwe West Indische Gids*, 69, 253–75.

Chamberlain, M. (2001), 'Migration, the Caribbean and the family', in Goulbourne, H. and Chamberlain, M. (eds), *Caribbean Families in Britain and the Trans-Atlantic World*, Oxford and London: Macmillan, 32–47.

Chamberlain, M. (2006), *Family Love in the Diaspora: Migration and the Anglo-Caribbean Experience*, New Brunswick, NJ: Transaction Publishers.

Christou, A. and King, R. (2006), 'Migrants encounter migrants in the city: the changing context of 'home' for second-generation Greek-American return migrants', *International Journal of Urban and Regional Research*, 30(4): 816–35.

Connell, J. (2003), 'An ocean of discontent? Contemporary migration and deprivation in the South Pacific', in Iredale, R., Hawksley, C. and Castles, S. (eds), *Migration in the Asia Pacific: Population, Settlement and Citizenship Issues*, Cheltenham, UK and Northampton, MA: Edward Elgar, 55–77.

Connell, J. and Conway, D. (2000), 'Migration and remittances in island microstates: a comparative perspective on the South Pacific and the Caribbean', *International Journal of Urban and Regional Research*, 24(1): 52–78.

Conway, D. (1985), 'Remittance impacts on development in the eastern Caribbean', *Bulletin of Eastern Caribbean Affairs*, 11: 31–40.

Conway, D. (1993), 'Rethinking the consequences of remittances for eastern Caribbean development', *Caribbean Geography*, 4: 116–30.

Conway, D. (2006), 'Globalization of labor: increasing complexity, more unruly', in Conway, D. and Heynen, N. (eds), *Globalization's Contradictions: Geographies of Discipline, Destruction and Transformation*, London: Routledge, 79–94.

Conway, D., Ellis, M. and Shiwdhan, N. (1990), 'Caribbean international circulation: are Puerto Rican women "tied-circulators"?' *Geoforum*, 21(1): 51–66.

Conway, D. and Potter, R.B. (2007), 'Caribbean transnational return migrants as agents of change', *Blackwell Geography Compass* [online early October 2006], 1(1): 25–45.

Dawson, L.R. (2007), 'Brain drain, brain circulation, remittances and development for the Caribbean', *The Caribbean Papers, No 2*, Waterloo, Ontario, Canada: The Centre for International Governance Innovation (CIGI). http://www.cigionline.org.

De Souza, R.M. (1998), 'The spell of the Cascadura: West Indian return migration', in Klak T (ed), *Globalisation and Neoliberalism: the Caribbean Context*, London: Rowman and Littlefield, 227–53.

Duany, J. (2002), *Puerto Rican Nation on the Move: Identities on the Island and in the United States*, Chapel Hill, NC and London: University of North Carolina Press.

Duval, D.T. (2002), 'The return visit-return migration connection', in Williams, A.M. and Hall, C.M. (eds), *Tourism and Migration: New Relationships between Production and Consumption*, Dordrecht, Boston and London: Kluwer Academic, 257–76.

Duval, D.T. (2004), 'Linking return visits and return migration among Commonwealth Eastern Caribbean migrants in Toronto', *Global Networks*, 4(1): 51–67.

Ellis, M., Conway, D. and Bailey, A.J. (1996), 'The circular migration of Puerto Rican women: towards a gendered explanation', *International Migration*, 34(1): 31–64.

Ghosh, B. (ed.), *Return Migration: Journey of Hope or Despair?*, Geneva: International Organization for Migration, United Nations.

Gmelch, G. (1987), 'Work, innovation and investment: the impact of return migrants in Barbados', *Human Organization*, 46, 131–140.

Gmelch, G. (1992), *Double Passage: The Lives of Caribbean Migrants Abroad and Back Home*, Ann Arbor, MI: University of Michigan Press.

Gmelch, G. (2006), 'Barbadian migrants abroad and back home', in Plaza, D.E. and Henry, F. (eds), *Returning to the Source: The Final Stage of the Caribbean Migration Circuit*, Jamaica, Barbados and Trinidad and Tobago: University of the West Indies Press, 49–73.

Goulbourne, H. (2002), *Caribbean Transnational Experience*, London: Pluto Press and Kingston: Arawak Publications.

Goulbourne, H. and Chamberlain, M. (eds), *Caribbean Families in Britain and the Trans-Atlantic World*, Oxford and London: Macmillan.

Ho, E.S. (2002), 'Multi-local residence, transnational networks: Chinese "astronaut" families in New Zealand', *Asian and Pacific Migration Journal*, 11(1): 145–64.

Hugo, G. (2003), 'Circular migration: keeping development rolling?', Migration Information Source, Washington DC: Migration Policy Institute, http://www.migrationinformation.org/feature/print.cfm?ID=129.

Kapur, D. and McHale, J. (2005), *Give Us Your Brightest and Your Best: The Global Hunt for Talent and its Impact on the Developing World*, Washington, DC: Center for Global Development.

King, R. (2000), 'Generalizations from the history of return migration', in Ghosh, B. (ed.), *Return Migration: Journey of Hope or Despair?*, Geneva: International Organization for Migration, United Nations, 7–55.

Laoire, C.N. (2008), '"Settling back"? A biographical and life-course perspective on Ireland's recent return migration', *Irish Geography*, 41(2): 195–210.

Lidgard, J. and Gilson, C. (2002), 'Return migration of New Zealanders: shuttle and circular migrants', *New Zealand Population Review*, 28(1): 99– 128.

Massey, D. (1984), 'Introduction: geography matters', in Massey, D. and Allen, J. (eds), *Geography Matters! A Reader*, London, New York, New Rochelle, Melbourne, Sydney: Cambridge University Press in association with The Open University, 1–11.

Papademetriou, D.G. (2003), 'Managing rapid and deep change in the newest age of migration', *The Political Quarterly*, 74(1). 39–58.

Plaza, D.E. (2002), 'The socio-cultural adjustments of second-generation British-Caribbean "return" migrants to Barbados and Jamaica', *Journal of Eastern Caribbean Studies*, 27: 135–60.

Plaza, D.E. and Henry, F. (2006), *Returning to the Source: The Final Stage of the Caribbean Migration Circuit*, Jamaica, Barbados and Trinidad and Tobago: University of the West Indies Press.

Portes, A. and Guarnizo, L.E. (1991), 'Tropical capitalists: US-bound immigration and small-enterprise development in the Dominican Republic', in Diaz-Briquets, S. and Weintraub, S. (eds), *Migration, Remittances and Small Business Development: Mexico and Caribbean Basin Countries*, Boulder: Westview, 101– 31.

Potter, R.B. (2001b), '"Tales of two societies": young return migrants to St Lucia and Barbados', *Caribbean Geography*, 12: 24–43.

Potter, R.B. (2001a), 'Narratives of socio-cultural adjustment among young return migrants to St Lucia and Barbados', *Caribbean Geography*, 12: 70–89.

Potter, R.B. (2005b), '"Citizens of descent": foreign-born and young returning nationals to St Lucia', *Journal of Eastern Caribbean Studies*, 30: 1–30.

Potter, R.B. (2005a), '"Young, gifted and back": second-generation transnational return migrants to the Caribbean', *Progress in Development Studies*, 5: 213–36.

Potter, R.B., Barker, D., Conway, D. and Klak, T. (2004), *The Contemporary Caribbean*, London and New York: Pearson/Prentice Hall.

Potter, R.B. and Conway, D. (2008), 'The development potential of Caribbean young return migrants: "making a difference back home ..."', in van Naerssen, T., Spaan, E. and Zoomers, A. (eds), *Global Migration and Development*, Abingdon, Oxford: Routledge, 213–30.

Pratt, G. and Yeoh, B. (2003), 'Transnational (counter), topographies', *Gender, Place and Space*, 10: 159–66.

Vertovec, S. (2007), 'Circular migration: the way forward in global policy?', Oxford: International Migration Institute, Working Paper, No. 4.

Wilson, P.J. (1969), 'Reputation and respectability: a suggestion for Caribbean ethnology', *Man*: 70–84

Index